Nick Vandome

MacBook

in easy steps

For MacBook, MacBook Air and MacBook Pro

7th Edition
Updated for macOS Big Sur (v11)

In easy steps is an imprint of In Easy Steps Limited
16 Hamilton Terrace · Holly Walk · Leamington Spa
Warwickshire · United Kingdom · CV32 4LY
www.ineasysteps.com

Seventh Edition

Notice of Liability

Every effort has been made to ensure that this book contains accurate
and current information. However, In Easy Steps Limited and the
author shall not be liable for any loss or damage suffered by readers
as a result of any information contained herein.

Trademarks

OS X, macOS, MacBook, MacBook Air and MacBook Pro are registered
trademarks of Apple Computer, Inc. All other trademarks are
acknowledged as belonging to their respective companies.

In Easy Steps Limited supports The Forest Stewardship Council (FSC),
the leading international forest certification organization. All our titles
that are printed on Greenpeace approved FSC certified paper carry the
FSC logo.

MIX
Paper from
responsible sources
FSC® C020837

Printed and bound in the United Kingdom

ISBN 978-1-84078-917-1

Contents

1 Introducing MacBooks

Apple's MacBook range of laptop computers is stylish and user-friendly. This chapter introduces the MacBook range so that you can choose the best one for your mobile computing needs.

About MacBooks

When Apple Computer, Inc. (renamed Apple Inc. in 1997) introduced its iMac range of desktop computers in 1998 it was a major breakthrough. To try to match the success of the iMac, Apple began working on a new range of notebook computers. It first entered this market seriously with the Macintosh Portable in 1989. In 1991, Apple introduced the PowerBook range of laptops, which was the forerunner to the MacBook range.

In 1999, a new range of Apple laptops was introduced. This was the iBook range, aimed firmly at the consumer market. In May 2006, the MacBook range first appeared. The two main reasons for this consolidation were:

- Simplifying Apple's laptop range under one banner.

- It was during this period that Apple Inc. was moving from PowerPC processors for its computers, to Intel processors.

The MacBook range now consists of:

- **MacBook Pro 13-inch**. The latest MacBook Pro modes come with two different display screen sizes (measured diagonally). One is a 13-inch Retina display screen model and is designed to be as thin and as light as possible. It also has an innovative trackpad with Force Touch technology that provides extra functionality.

- **MacBook Pro 16-inch**. This is the most powerful version of the MacBook, and like the 13-inch models contains the Touch Bar, for a range of added functionality from the keyboard.

- **MacBook Air**. This range was designed to be the thinnest and lightest on the market, and it is still an ultraportable laptop and ideal for mobile computing.

The New icon indicates a new or enhanced feature introduced with the latest version of macOS Big Sur on the MacBook.

The MacBook Pro 13-inch and the MacBook Air are the first models to use Apple's own M1 processor chip, as opposed to Intel ones.

MacBook Models

Specifications for all computers change rapidly, and for the current MacBook range they are (at the time of printing):

MacBook Pro 13-inch
The specifications for this model are:

- Processor: Apple M1 chip, or 2.0GHZ quad-core Intel Core i5 with 6MB shared L3 cache.

- Storage: 256 gigabytes (GB) or 512GB, both configurable up to 2 terabytes (TB).

- Memory: 8GB, configurable to 16GB.

- Ports: Two Thunderbolt/USB 4 ports, with support for USB 3.1.

- Battery: Up to 17 hours of wireless web use.

MacBook Pro 16-inch
The specifications for this model are:

- Processor: 2.6GHZ 6-core Intel Core i7, or 2.3GHz 8-core Intel Core i9, with 12MB/16MB shared L3 cache.

- Storage: 512GB or 1TB.

- Memory: 16GB, configurable to 32GB or 64GB.

- Ports: Four Thunderbolt 3/USB-C ports, with support for USB 3.1.

- Battery: Up to 11 hours of wireless web use.

MacBook Air
The specifications for this model are:

- Processor: Apple M1 chip.

- Storage: 256GB or 512GB.

- Memory: 8GB, configurable to 16GB.

- Ports: Two Thunderbolt/USB 4 ports, with support for USB 3.1.

- Battery: Up to 15 hours of wireless web use.

All MacBooks have a range of energy-saving and environmental features.

The storage and memory on the MacBook range can both be configured to higher levels.

The MacBook Air and the MacBook Pro both have flash storage. This is similar in some ways to traditional ROM (read-only memory) storage, but it generally works faster and results in improved performance.

MacBook Jargon Explained

Since MacBooks are essentially portable computers, a lot of the jargon is the same as for other computers. However, it is worth looking at some of this jargon and the significance it has in terms of MacBooks.

- **Processor**. Also known as the central processing unit, or CPU, this refers to the processing of digital data as it is provided by apps on the computer. The more powerful the processor, the quicker the data is interpreted. Apple is in the process of moving its processors from Intel to its own M1 range, which is included in the MacBook Pro 13-inch and the MacBook Air.

- **Memory**. This closely relates to the processor and is also known as random-access memory, or RAM. Essentially, this type of memory manages the apps that are being run and the commands that are being executed. The greater the amount of memory there is, the quicker the apps will run. With more RAM they will also be more stable and less likely to crash. In the current range of MacBooks, memory is measured in gigabytes (GB) and ranges from 8GB to 64GB.

- **Storage**. This refers to the amount of digital information the MacBook can store. It is frequently referred to in terms of hard disk space and is measured in gigabytes (GB) or terabytes (TB). MacBooks use Solid-State Drive (SSD) flash storage.

- **Trackpad**. This is an input device that takes the place of a mouse (although a mouse can still be used with a MacBook, either with a cable or wirelessly). Traditionally, trackpads have come with a button that duplicates the function of the buttons on a mouse. However, the trackpad on a MacBook has no button, as the pad itself performs these functions.

Memory can be thought of as a temporary storage device, as it only keeps information about the currently-open apps. Storage is more permanent, as it keeps the information even when the MacBook has been turned off.

All of the most recent MacBooks have the Force Touch trackpad, which provides additional functionality for pressing on the trackpad.

● **Graphics card**. This is a device that enables images, videos and animations to be displayed on the MacBook. It is also sometimes known as a video card. The faster the graphics card, the better the quality relevant media will be displayed at. In general, very fast graphics cards are really only needed for intensive multimedia applications, such as video games or videos.

● **Wireless**. This refers to a MacBook's ability to connect wirelessly to a network; i.e. another computer or an internet connection. In order to be able to do this, the MacBook must have a wireless card, which enables it to connect to a network or high-speed internet connection. This is known as the AirPort Extreme Wi-Fi wireless networking card.

● **Bluetooth**. This is a radio technology for connecting devices wirelessly over short distances. It can be used for items such as a wireless mouse or for connecting to a device, such as an iPhone for downloading photos.

● **Ports**. These are the parts of a MacBook that external devices can be plugged into, using a cable that connects into one of the ports, which are located on the side of the MacBook. The latest range of MacBooks use a headphone port for attaching a pair of headphones for listening to music, podcasts or video content such as movies or TV shows.

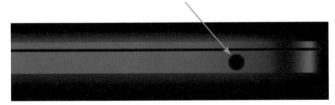

For attaching other external devices, MacBooks use Thunderbolt ports. These transfer data at high speeds. and can also be used to connect a range of USB devices (see page 12).

Another option for connecting to a network is using an Ethernet cable, rather than doing it wirelessly. The latest range of MacBooks do not have a dedicated Ethernet port, but a Thunderbolt to Ethernet adapter can be used.

Thunderbolt ports can also be used to attach a Thunderbolt Display monitor.

...cont'd

- **USB**. This is a method for connecting a variety of external devices, such as digital cameras, scanners, and printers. The latest range of MacBooks do not have separate USB ports, but the Thunderbolt ports can accommodate USB 4 devices. They also support older versions of USB devices (USB 3.1 and earlier), but an adapter is required to connect earlier versions of USB devices to the Thunderbolt port.

Don't forget

USB stands for Universal Serial Bus. The type of adapter required for attaching USB 3.1 and earlier devices to a Thunderbolt port is a USB-C to USB adapter.

- **Touch Bar**. This is available on the 13-inch and 16-inch models of the MacBook Pro. It is a multi-touch strip located at the top of the keyboard that offers a range of options, depending on the current app being used; e.g. the system controls, such as for adjusting volume and screen brightness, or adding emojis to an iMessage.

Don't forget

For more details about the Touch Bar, see pages 26-27.

- **CD/DVD players or re-writers**. The latest range of MacBooks do not have a built-in CD/DVD player  or re-writer. However, an external USB SuperDrive can be purchased for playing DVDs and CDs or burning content to a CD or DVD. For copying content, flashdrives are also a popular option as they can contain large amounts of data and connect via a USB port.

Don't forget

When a SuperDrive is attached to a MacBook, it shows up as an external drive in the Finder; see page 32.

- **Webcam (FaceTime)**. This is a type of camera fitted into the MacBook and it can be used to take still photographs or communicate via video with other people. On the MacBook it is known as the FaceTime camera, and it works with the FaceTime app. The FaceTime camera is built in at the top middle of the inner casing.

Getting Comfortable

Since you will probably be using your MacBook in more than one location, the issue of finding a comfortable working position can be vital, particularly as you cannot put the keyboard and monitor in different positions, as you can with a desktop computer. Whenever you are using your MacBook, try to make sure that you are sitting in a comfortable position with your back well supported, and that the MacBook is in a position where you can reach the keyboard easily, and also see the screen, without straining your arms.

Despite the possible temptation to do so, avoid using your MacBook in bed, on your lap, or where you have to slouch or strain to reach the MacBook properly.

Seating position
The ideal way to sit at a MacBook is with an office-type chair that offers good support for your back. Even with these types of chairs, it is important to maintain a good body position so that your back is straight and your head is pointing forwards.

If you do not have an office-type chair, use a chair with a straight back and place a cushion behind you for extra support and comfort, as required.

If possible, the best place to work on a MacBook is at a dedicated desk or workstation.

One of the advantages of office-type chairs is that the height can usually be adjusted, and this can be a great help in achieving a comfortable position.

...cont'd

MacBook position

When working at your MacBook it is important to have it positioned so that both the keyboard and the screen are in a comfortable position. If the keyboard is too low, you will have to slouch or strain to reach it.

If the keyboard is too high, your arms will be stretching. This could lead to pain in your tendons.

The ideal setup is to have the MacBook in a position where you can sit with your forearms and wrists as level as possible while you are typing on the keyboard.

Adjusting the screen

Another factor in working comfortably at a MacBook is the position of the screen. Unlike a desktop computer, it is not feasible to have a MacBook screen at eye level, as this would result in the keyboard being in too high a position. Instead, once you have achieved a comfortable seating position, open the screen so that it is approximately 90 degrees from your eyeline.

Working comfortably at a MacBook involves a combination of a good chair, good posture and good MacBook positioning.

One potential issue with MacBook screens can be that they reflect glare from sunlight or indoor lighting.

If this happens, either change your position or block out the light source, using some form of blind or shade. Avoid squinting at a screen that is reflecting glare, as this will make you feel uncomfortable and quickly give you a headache.

Input Devices

MacBooks have the same data input devices as most laptops: a keyboard and a trackpad. However, the trackpad has an innovative feature that makes it stand out from the crowd: there is no button – the trackpad itself performs the functions of a button. The trackpad uses Multi-Touch Gestures to replace traditional navigation techniques. These are looked at in detail in Chapter 6. Some of these gestures are:

- **One-finger click**. Click in the middle of the trackpad to perform one-click operations.

- **Scrolling**. This can be done on a page by dragging two fingers on the trackpad either up or down.

- **Zooming on a page or web page**. This can be done by double-tapping with two fingers.

When using the keyboard or trackpad, keep your hands and fingers as flat as possible over the keyboard and trackpad.

Trackpad options

Options for the functioning of the trackpad can be set within **System Preferences** (see pages 28-30). To do this:

1 Access **System Preferences** and click on the **Trackpad** button

Trackpad

2 Click on the tabs to set options for pointing and clicking, scrolling and zooming, and additional Multi-Touch Gestures with the trackpad

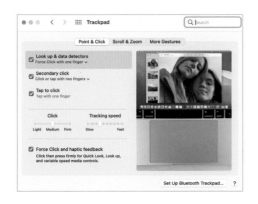

Don't forget

The trackpad on a MacBook is quite sensitive, so only a relatively small amount of pressure is required for Multi-Touch Gestures.

…cont'd

Mouse options

An external mouse can be connected to a MacBook, and options for its functioning can be set within **System Preferences**. To do this:

1 Click on the **Mouse** button

Mouse

2 Drag the sliders to set the speed at which the cursor moves across the screen, and also the speed required for a double-click operation

16

Keyboard options

Options for the functioning of the keyboard can be set within **System Preferences**. To do this:

1 Click on the **Keyboard** button

Keyboard

2 Click on the **Keyboard** tab to set options for how the keyboard operates, such as the speed for repeating a key stroke

3 Click on the **Text** tab to set keyboard shortcuts for accessing certain words and phrases. Click on this button to add a new shortcut and phrase

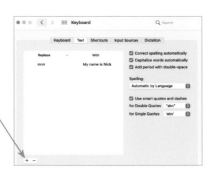

MacBook Power Cable

All MacBooks need a power cable – connected to an AC/DC adapter – that can be used to recharge the battery, and it can also be used when the MacBook is not being used in a mobile environment. This can save the battery so, if possible, the adapter should be used instead of battery power.

The latest range of MacBooks use USB-C power adapters, which connect to the Thunderbolt port with a USB-C cable. One end of the cable attaches to the power adapter and the other end attaches to the Thunderbolt port on the MacBook. The power adapter and the USB-C cable come supplied with the MacBook.

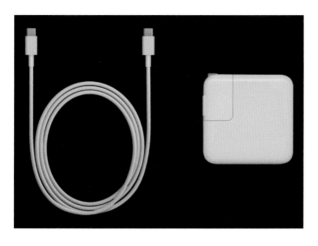

For more information about accessing and using System Preferences, see pages 28-30.

Power management

Within the Battery system preferences there are options for when the MacBook is connected to the power adapter (as well as when it is using battery power). Click on the **Power Adapter** tab in the left-hand sidebar and make the appropriate selections.

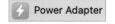

For more information about accessing and using the Battery system preferences, see pages 167-169.

Cleaning a MacBook

Like most things, MacBooks benefit greatly from a little care and attention. The two most important areas to keep clean are the screen and the keyboard.

Cleaning the screen

All computer screens quickly collect dust and fingerprints, and MacBooks are no different. If this is left too long, it can make the screen harder to read and cause eye strain and headaches. Clean the screen regularly with the following cleaning materials:

- A lint-free cloth, similar to the type used to clean camera lenses (it is important not to scratch the screen in any way).

- An alcohol-free cleaning fluid recommended for computer screens.

- Screen wipes that are recommended for use on computer screens.

Cleaning the keyboard

Keyboards are notorious for accumulating dust, fluff and crumbs. One way to solve this problem is to turn the MacBook upside down and very gently shake it to loosen any foreign objects. Failing this, a can of compressed air can be used, with a narrow nozzle to blow out any stubborn items that remain lodged between the keys.

The outer casing of a MacBook can be cleaned with the same fluid as used for the screen. A duster or a damp (but not wet) cloth and warm water can be equally effective. Keep soap away from MacBooks if possible.

Spares and Accessories

Whenever you are going anywhere with your MacBook, there are always spares and accessories to consider. Some of these are just nice things to have, while others could be essential in ensuring that you can still use your MacBook if anything goes wrong while you are on your travels. Items to consider putting in your MacBook case, or using at home, include:

Hardware

- **Replacement battery**. Although it is technically possible to change the battery of a MacBook, it is not recommended, as it could damage the device and would invalidate any warranty. MacBook batteries are designed to retain up to 80% of their original capacity for 1,000 complete charge cycles. However, if a MacBook battery does die, it can be replaced by using Apple's battery replacement service: details can be found on the Apple website at **https://support.apple.com/mac/repair/service**, or at your nearest Apple Store.

- **Apple Thunderbolt Display**. The Thunderbolt port enables you to transfer data from peripheral devices to your MacBook at very high speeds. You can also connect a Thunderbolt Display – a 27-inch display that can give stunning clarity to the content on your monitor.

- **Magic Mouse**. This is an external mouse that can be connected to your MacBook and used to perform Multi-Touch Gestures in the same way as with the trackpad.

- **External SuperDrive**. The MacBook range does not come with a SuperDrive for using CDs and DVDs but this external one can be used to connect to a MacBook using a Thunderbolt cable, or a USB one with a Thunderbolt adapter.

- **Wireless keyboard**. This can be used if you want to have a keyboard that you can move away from your MacBook.

- **Thunderbolt Ethernet adapter**. This can be used to connect to an Ethernet network using an Ethernet cable.

- **Multi-card reader**. This is a device that can be used to copy data from the cards used in digital cameras. If you have a digital camera, it is possible to download the photographs from it directly onto a MacBook with a cable. However, a multi-card reader can be more efficient and flexible.

Beware

If a MacBook is still within its one-year warranty, a defective battery will be replaced for free, as long as it has not been damaged due to misuse.

...cont'd

Hot tip

Apple's AirPods are a good option for listening to items on your MacBook, as they are wireless and can be connected with Bluetooth.

20

Don't forget

The App Store can be accessed by clicking on this button on the Dock at the bottom of the MacBook screen:

- **Headphones/EarPods**. These can be used to listen to music or movies if you are in the company of other people and you do not want to disturb them. They can also be very useful if there are distracting noises from other people.

- **Flashdrive**. This is a small device that can be used to copy data to and from your MacBook. It connects via a Thunderbolt/USB port and is about the size of a packet of chewing gum. It is an excellent way of backing up files from your MacBook when you are away from home or the office.

- **Cleaning material**. The materials described on page 18 can be taken to ensure your MacBook is always in tip-top condition for use.

- **CDs/DVDs**. Video or music CDs and DVDs can be taken to provide mobile entertainment, and blank ones can be taken to copy data onto, similar to using a pen drive. An external CD/DVD drive will also be required.

Software

New software programs, or apps, can be downloaded directly from the Apple App Store. There is a huge range on offer, covering over 20 different categories. They can be accessed from the left-hand sidebar and you can also view and update apps that you have purchased (see pages 133-138).

2 Around a MacBook

This chapter looks at getting started with your MacBook: from opening it up and turning it on, through to keyboard functions and System Preferences. It also looks at the Touch Bar.

Opening Up

The first step toward getting started with a new MacBook is to open it ready for use.

There is no physical latch on the front of a MacBook; just a small groove at the front where the top and bottom parts meet.

Instead of a latch, a MacBook has a magnetic closing mechanism, which engages when the monitor screen is closed onto the main body of the MacBook. To open it, raise it firmly upwards from the center, where there is a smooth groove.

Once the MacBook has been opened, the screen should stay in whatever position it is placed.

Open the screen of a MacBook carefully, so as not to put any unnecessary pressure on the connection between the screen and the main body of the MacBook.

To turn on the MacBook, press the button in the top right-hand corner of the keyboard.

On models with the Touch Bar, this button is located at the right-hand side of the bar.

MacBook Desktop

The opening view of a MacBook is known as the Desktop. Items such as apps and files can be stored on the Desktop but, in general, it is best to try to keep it as clear as possible.

At the top of the Desktop are the Apple menu (Apple symbol) and the Menu bar. This contains links to a collection of commonly-used menus and functions, such as Copy and Paste.

The menus on the Menu bar are looked at in detail on page 100.

At the bottom of the Desktop is the Dock. This is a collection of icons that are shortcuts to frequently-used apps or folders.

One of the items on the Dock is the Finder. This can be used to access the main area for apps, folders and files, and also to organize the way you work on your MacBook.

To specify which items appear on the Desktop, click on **Finder** > **Preferences** on the Menu bar. Click on the **General** tab and select to show or hide **Hard disks**, **External disks**, **CDs, DVDs and iPods** or **Connected servers**. The selected items will then be displayed as icons on the Desktop.

Apple Menu

The Apple menu is accessed from the Apple symbol at the left-hand side of the Menu bar:

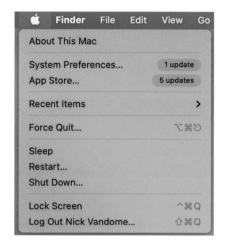

Hot tip

Force Quit can be used to close down an app that is frozen or is not responding.

The options on the Apple menu are:

- **About This Mac**. This provides general information about the processor, the amount of memory and the Big Sur version.

- **System Preferences**. This is a shortcut to System Preferences. This can be used to access a wide range of options, including those for items such as the Dock.

- **App Store**. This can be used to access the online Apple App Store for downloading apps, and also software updates for existing apps and Big Sur.

- **Recent Items**. This displays the items you have most recently used and viewed.

- **Force Quit**. This can be used to manually quit an app that has frozen or will not close.

- **Sleep**. This puts the MacBook into a state of hibernation.

- **Restart**. This closes down the MacBook and restarts it.

- **Shut Down**. This shuts down the MacBook.

- **Log Out**. This shuts down the currently-open apps, and logs out the current user.

Standard Keyboard Buttons

A MacBook keyboard has a number of keys that can be used for shortcuts or specific functions. Four of them are located at the left of the space bar. They are (from left to right):

- **The Function key**. This can be used to activate the function (Fn) keys at the top of the keyboard: press the Fn key and the required F key at the same time.

- **The Control key**. This can be used to access contextual menus.

- **The Alt (Option) key**. This is frequently used in conjunction with the Command key to perform specific tasks.

- **The Command key**. As above.

At the top of the keyboard there are F keys for changing some of the settings on your MacBook. These are (from left to right):

- F1: Decrease brightness.

- F2: Increase brightness.

- F3: Show/Hide all open windows (Mission Control).

- F4: Show/Hide the Launchpad function for displaying all of the apps on your MacBook.

- F5/F6: Adjust the keyboard brightness up or down.

- F7: Rewind a music track or video.

- F8: Play/Pause a music track or video.

- F9: Fast-forward a music track or video.

- F10: Mute volume.

- F11: Decrease volume (with Fn key, displays the Desktop and minimizes all windows around the sides of the screen).

- F12: Increase volume.

Contextual menus are ones that have actions specific to the item being viewed.

On models of MacBook with the Touch Bar, the F keys appear on the Touch Bar rather than physically on the keyboard, and they are accessed from the Fn key on the keyboard.

Touch Bar

The Touch Bar is an innovation that is included with the 13-inch and 16-inch models of the MacBook Pro. It is a glass strip at the top of the keyboard that offers multi-touch dynamic functionality, tailored to the app being used. It operates in a similar way to the screen of a smartphone, such as the iPhone, in that options can be accessed by tapping and swiping on it. The Touch Bar also includes Touch ID functionality (at the far right-hand side of the Touch Bar), which can be used to unlock the MacBook using a fingerprint (in the same way as it is used on an iPhone and an iPad).

Hot tip

The area on the far right-hand side of the Touch Bar can be used to turn on the MacBook, unlock it with Touch ID, and also pay for online items with Apple Pay at participating retailers.

Using the Touch Bar

The Touch Bar options change according to which app is currently being used. Also, some apps have additional options that can be accessed by swiping from right to left on the Touch Bar.

Don't forget

More apps are being created by third-party developers that support Touch Bar functionality so, in time, there will be more and more apps that can be used with the Touch Bar.

Although the options on the Touch Bar change with different apps, there are some default buttons that appear on some of the Touch Bar options, but not all of them. These are for amending brightness and volume, and accessing Siri.

Tap on the left-pointing arrow to expand these options.

Touch Bar options

Some of the Touch Bar options, for specific apps, include:

General Settings: This can be used to change screen brightness and contrast, access Mission Control for viewing open apps, use video controls and adjust volume.

Web pages – Safari app: This can be used to move between the previous and next web pages, search the web, access different tabs in Safari, view tabs and add new tabs.

Email – Mail app: This can be used to create, format and send email messages.

Text messages – Messages app: This can be used to add a range of emojis to text messages.

Color selection – Keynote, Pages, Photoshop and other apps with color selection: This can be used to make color selections with a slider or color wheel for apps that have this option.

F keys: The traditional F keys are available on the Touch Bar by pressing the Fn key on the keyboard.

System Preferences

In macOS there are preferences that can be set for just about every aspect of your computer. This gives you greater control over how the interface looks and how the operating system functions. To access System Preferences:

1 Click on this icon on the Dock or from the Applications folder in the Finder

Apple ID. This contains options for setting up an Apple ID for use with iCloud, the online sharing and storage service.

Family Sharing. This can be used to set up and use Family Sharing, for sharing music, movies, games, apps and books with up to five other family members.

General. Options for the overall look of buttons, menus, windows and scroll bars.

Desktop & Screen Saver. This can be used to change the Desktop background and the screen saver.

Dock & Menu Bar. Options for the way the Dock and the top Menu bar look and operate.

Mission Control. This gives you a variety of options for managing all of your open windows and apps.

Siri. This can be used to turn on or off the digital voice assistant, Siri, and also set voice style and language.

Spotlight. This can be used to specify settings for the macOS search facility, Spotlight.

28

Hot tip

The **Dock & Menu** section also includes options for what appears in the Control Center, which is accessed from this icon on the Menu bar. See pages 64-65 for details about accessing and using the Control Center.

Language & Region. Options for the language used on your MacBook.

Notifications. This can be used to set up how you are notified about items such as calendar, messages and software updates.

Internet Accounts. This can be used to link to other online accounts that you have (see Hot tip).

Users & Groups. This can be used to create accounts for different users on your MacBook.

Accessibility. This can be used to set options for users who have difficulty with viewing text on screen, hearing commands, using the keyboard or using the mouse.

Screen Time. This can be used to show screen usage and limit access to the computer and various online functions.

Hot tip

The **Internet Accounts** section can be used to set up online email accounts such as Google or Yahoo.

...cont'd

Extensions. This can be used to customize your MacBook with extensions and plug-ins from Apple and third-party developers.

Security & Privacy. This enables you to secure your Home folder with a master password, for added security.

Software Update. This can be used to specify how software updates are handled. It connects to the App Store to access the available updates.

Network. This can be used to specify network settings for connecting to the internet or other computers.

Bluetooth. Options for attaching Bluetooth wireless devices.

Sound. Options for adding sound effects, and playing and recording sound.

Printers & Scanners. Options for selecting printers and scanners.

Keyboard. Options for how the keyboard functions; your keyboard shortcuts; and dictation settings.

Trackpad. Options for when you are using a trackpad.

Mouse. Options for how the mouse functions.

Displays. Options for the screen display, such as resolution.

Sidecar. This can be used to connect to an iPad, to use it as an additional monitor for a Mac.

Battery. Options for managing the MacBook's battery. See pages 166-172 for details about managing the battery.

Date & Time. Options for changing the computer's date and time to time zones around the world.

Sharing. This can be used to specify how files are shared over a network. This is also covered on page 155.

Time Machine. This can be used to configure and set up the macOS backup facility. See pages 174-177 for details about setting up and using Time Machine.

Startup Disk. This can be used to specify the disk from which your computer starts up. This is usually the macOS volume.

Don't forget

macOS Big Sur supports multiple displays, which means you can connect your MacBook to two or more displays and view different content on each one. The Dock appears on the active screen, and each screen also has its own Menu bar. Different full-screen apps can also be viewed on each screen.

Connecting a Printer

Using a printer on any computer is essential, and MacBooks allow you to quickly add a printer to aid your productivity. To do this:

1 Open **System Preferences** and click on the **Printers & Scanners** button

2 Currently-installed printers are displayed in the **Printers** list

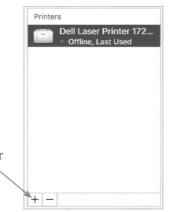

3 Click here to add a new printer or any currently-available printers

4 Select a printer and click on the **Add** button

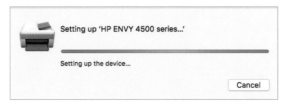

5 macOS Big Sur loads the required printer driver. (If it does not have a specific one it will try to use a generic one)

6 Details about the printer are available in the **Printers** window

Printer drivers are programs that enable the printer to communicate with your computer. Printer drivers are usually provided on a disc when the printer is purchased. In addition, MacBooks will have a number of printer drivers pre-installed. If your MacBook does not recognize your printer, you can load the driver from the disc.

Once a printer has been installed, documents can be printed by selecting **File** > **Print** from the Menu bar. Print settings can be set at this point, and they can also be set by selecting **File** > **Page/Print Setup** from the Menu bar in most apps.

External Drives

Attaching external drives is an essential part of mobile computing, whether it is to back up data as you are traveling, or for downloading photos and other items. On MacBooks, external drives are displayed on the Desktop once they have been attached, and they can then be used for the required task. To do this:

Hot tip

If external drives do not appear on the Desktop, select **Finder** > **Preferences** from the Menu bar. Click on the **General** tab and select to show or hide **External disks**.

1 Attach the external drive. This is usually done with a Thunderbolt/USB cable. Once it has been attached, it is shown on the Desktop

2 The drive is shown in the Finder

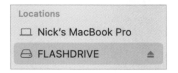

Don't forget

External drives can be items such as flashdrives, digital cameras or external hard disks.

3 Perform the required task for the external drive (such as copying files or folders onto it from the hard drive of your MacBook, using the Finder)

Don't forget

An external drive can be renamed by **Ctrl** + **clicking** on its name in the Finder and overtyping it with a new name.

4 External drives should be ejected properly, not just pulled out or removed. To do this, click on this button next to the drive in the Finder window, or drag its icon on the Desktop over the **Trash** icon on the Dock. This will then change into an **Eject** icon, and the drive can then be removed

3 Introducing macOS Big Sur

Big Sur is the latest operating system for MacBooks. This chapter introduces some of its essential features so that you can quickly feel comfortable using it. It also covers the digital voice assistant, Siri, which is included on MacBooks.

About macOS Big Sur

macOS Big Sur 11 is the latest version of the operating system for Apple computers, including the MacBook range. It is the first version of the operating system to have the 11 designation, after over a decade of versions with the 10 designation. This is to recognize the new interface for macOS and also the fact that new Mac desktops and laptops now use ARM-based processors, rather than Intel ones. macOS is still based on the UNIX programming language, which is a very stable and secure operating environment, and this ensures that macOS is one of the most stable consumer operating systems that has ever been designed. It is also one of the most stylish and user-friendly operating systems available.

macOS Big Sur continues the functionality of several of its Mac operating system predecessors, including using some of the functionality that is available on Apple's mobile devices such as the iPhone and the iPad. The two main areas where the functionality of the mobile devices has been incorporated are:

- The way apps can be downloaded and installed. Instead of using a disc, macOS Big Sur utilizes the App Store to provide apps, which can be installed in a couple of steps.

- Using Multi-Touch Gestures (with a laptop trackpad, Apple's Magic Trackpad, or Magic Mouse) for navigating apps and web pages.

macOS Big Sur continues the evolution of the operating system, by adding features and enhancing existing ones. These include:

- The Control Center of useful widgets, which can be accessed from the top toolbar in any macOS Big Sur screen. This has been available on iPhones and iPads for a number of years.

- A redesigned Notifications Center that combines notifications and useful widgets for a range of tasks, such as obtaining weather forecasts, calendar events or stock prices.

- The sidebar in a number of macOS Big Sur apps has been standardized so that it can be displayed, or hidden, with a single click on this button on the app's toolbar.

- A number of apps have been updated in macOS Big Sur, including Safari, Messages, Maps and Notes.

macOS Big Sur is the latest version of the MacBook operating system.

UNIX is an operating system that has traditionally been used for large commercial mainframe computers. It is renowned for its stability and ability to be used within different computing environments.

Big Sur has a Power Nap function that updates items from the online iCloud service even when a MacBook is in Sleep mode. This can be set up by checking **On** the **Enable Power Nap** option in either the **Battery** section or the **Power Adapter** section of the **Battery** system preferences.

Installing macOS Big Sur

When it comes to installing macOS Big Sur you do not need to worry about an installation CD or DVD; it can be downloaded and installed directly from the online App Store. New MacBooks will have macOS Big Sur installed, and the following range is compatible with macOS Big Sur and can be upgraded with it:

- MacBook Pro (Late 2013 or newer).

- MacBook Air (Mid-2013 or newer).

If you want to install macOS Big Sur on an existing MacBook, you will need to have the minimum requirements of:

- 2GB of memory.

- 12.5GB of available storage for installation.

If your MacBook meets these requirements, you can download and install macOS Big Sur, for free, as follows:

macOS Big Sur is a new version of the MacBook operating system. It contains a number of new technologies to make the operating system faster and more reliable than ever.

1 Click on this icon on the Dock to access the App Store (or select **Software Update...**; see Hot tip)

2 Locate the **macOS Big Sur** icon using the App Store Search box

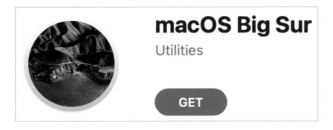

macOS Big Sur

Utilities

GET

3 Click on the **Get** button and follow the installation instructions GET

Hot tip

To check your computer's software version and upgrade options, click on **Apple menu** > **About This Mac** from the main Menu bar. Click on the **Overview** tab and click on the **Software Update...** button. See page 39 for details.

The Big Sur Environment

The first most noticeable element about Big Sur is its elegant user interface. This has been designed to create a user-friendly graphic overlay to the UNIX operating system at the heart of Big Sur, and it is a combination of rich colors and sharp, original graphics. The main elements that make up the initial Big Sur environment are:

Apple menu icon   Menu bar menus   Windows   Menu bar icons

The Dock   Desktop

The Dock is designed to help make organizing and opening items as quick and easy as possible. For a detailed look at the Dock, see pages 54-63.

Many of the behind-the-scenes features of macOS Big Sur are aimed at saving power on your MacBook. These include timer-coalescing technologies for saving processing and battery power; features for saving energy when apps are not being used; power-saving features in Safari for ignoring additional content provided by web page plug-ins; and memory compression to make your MacBook quicker and more responsive.

The **Apple menu** is standardized throughout Big Sur, regardless of the app in use.

Menus

Menus in Big Sur contain commands for the operating system and any relevant apps. If there is an arrow next to a command it means there are subsequent options for the item. Some menus also incorporate the same transparency as the sidebar so that the background shows through.

...cont'd

Transparency

One feature in macOS Big Sur is that the sidebar and toolbars in certain apps are transparent so that you can see some of the screen behind it. This also helps the uppermost window blend in with the background:

1 In certain apps with a sidebar, such as the Finder or the Safari sidebar, the background appears behind the sidebar

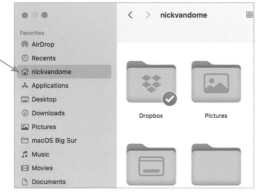

2 When you move the window, the background behind the sidebar changes accordingly

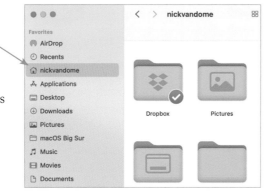

The red window button is used to close a window. However, this does not quit the app. The amber button is used to minimize a window, so that it appears at the right-hand side of the Dock.

37

Window buttons

These appear in any open macOS window and can be used to manipulate the window. They include a full-screen option. Use the window buttons to, from left to right: close a window, minimize a window or access full-screen mode (if available).

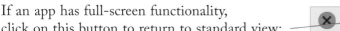

If an app has full-screen functionality, click on this button to return to standard view:

If an option is gray, it means it is not available; i.e. the screen cannot be maximized.

About Your MacBook

When you buy a new MacBook you will almost certainly check the technical specifications before you make a purchase. Once you have your MacBook, there will be times when you will want to view these specifications again, such as the version of macOS in use, the amount of memory, and the amount of storage. This can be done through the **About This Mac** option that can be accessed from the Apple menu. To do this:

1 Click on the **Apple menu** button and click on the **About This Mac** link

2 Click on the **Overview** tab

3 This window contains information about the version of macOS being used; processor; amount of memory; type of graphics card; and serial number

4 Click on the **System Report...** button to view full details about the hardware and software on your MacBook

System Report...

Hot tip

The **System Report** section is also where you can check whether your MacBook is compatible with the Handoff functionality (covered on page 75). Click on the **Bluetooth** section in **System Report** to see if Handoff is supported.

5 Click on the **Software Update...** button to see available software updates for your MacBook

Display information
This gives information about your MacBook's display:

1 Click on the **Displays** tab

Displays

2 This window contains information about your display including the type, size, resolution and graphics card

For more information about setting software updates, see page 181.

3 Click on the **Displays Preferences...**

Displays Preferences...

button to view options for changing the display's resolution, brightness and color

...cont'd

Storage information

This contains information about your MacBook's physical and removable storage:

1 Click on the **Storage** tab   Storage

2 This window contains information about the used and available storage on your hard disk, and also options for writing various types of CDs and DVDs

Managing memory

This contains information about your MacBook's memory, which is used to run macOS and also the applications on your computer:

1 Click on the **Manage...** tab   Manage...

2 This window contains information about how memory can be managed in macOS Big Sur (see tip)

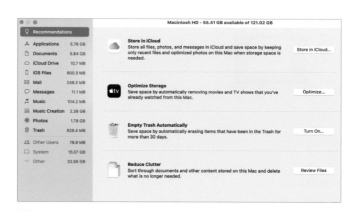

Don't forget

The options for managing storage include: **Store in iCloud**, which can be used to enable your MacBook to identify files that have not been opened or used in a long time and then automatically store them in iCloud (if you have enough storage space there); **Optimize Storage**, which can be used to automatically remove movies and TV shows, from the TV app, that have been watched; **Empty Trash Automatically**, which can automatically remove items from the Trash after 30 days; and **Reduce Clutter**, to delete content that is no longer needed.

Support

The **Support** tab provides links to a range of help options for your MacBook and macOS.

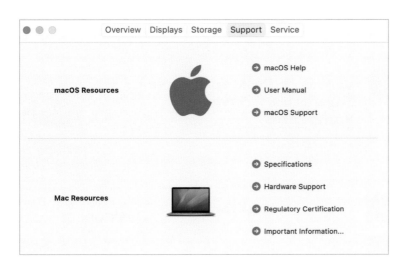

The **Support** options open in the relevant support section of the Apple website.

Service

The **Service** tab provides links to service and repair options and also the AppleCare Protection Plan, for extending the initial one-year warranty for your MacBook.

Customizing Your MacBook

Background imagery is an important way to add your own personal touch to your MacBook. (This is the graphical element upon which all other items on your computer sit.) There is a range of background options that can be used. To select your own background and screen saver:

Don't forget

You can select your own photographs as your Desktop background, once you have loaded them onto your MacBook. To do this, select one of the Photos folders in Step 3 and browse to the photograph you want.

Hot tip

Select one of the Dynamic Desktop options in Step 4 to create a background that changes according to the time of day; e.g. during the day it will be bright, and at night it will change to a darker version of the image.

1 Click on the **Desktop & Screen Saver** button in **System Preferences**

Desktop & Screen Saver

2 Click on the **Desktop** tab

Desktop

3 Select a location from where you want to select a background

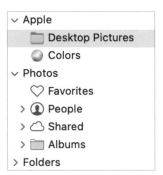

4 Click on one of the available backgrounds

5 The background is applied as the Desktop background imagery

...cont'd

6 Click on the **Screen Saver** tab

Screen Saver

7 Click here to
select a screen
saver, which
is previewed
in the main
window

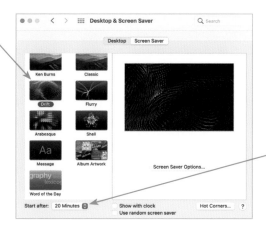

Click in the screen saver
Start after: box to
specify a period of time
of inactivity after which
the screen saver starts.

General customization

Other customization options can be found in the General section
of System Preferences:

1 Click on the **General** button in the
System Preferences folder

General

✓ Never
1 Minute
2 Minutes
5 Minutes
10 Minutes
20 Minutes

2 The General
window has
options for
items including
changing the
appearance of
buttons, menus
and windows;
changing
highlight
colors;
specifying
when scroll
bars appear;
and setting the
default web browser

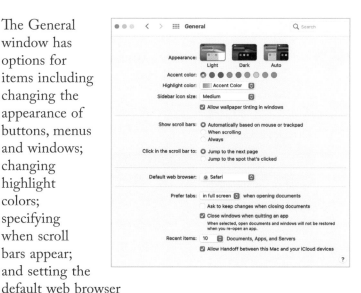

Search Options

Finding items on a MacBook or on the web is an important part of everyday computing, and with macOS Big Sur there are a number of options for doing this.

Siri

Siri is Apple's digital voice assistant that can search for items using speech. It has been available on Apple's mobile devices using iOS for a number of years, and with macOS Big Sur it is available on MacBooks. See pages 46-48 for more details about setting up and using Siri.

Spotlight search

Spotlight is a dedicated search function for macOS. It can be used over all the files on your MacBook and the internet. To use it:

1 Click on this icon at the right of the Apple Menu bar

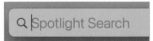

2 Click in the Spotlight Search box

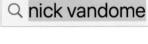

3 Enter a keyword, or phrase, for which you want to search

4 The top hits from within your apps, and the web, are shown in the left-hand panel. Click on an item to view its details

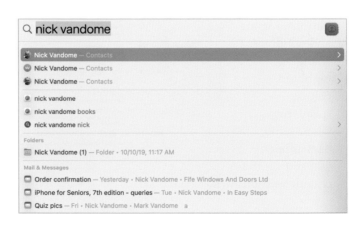

44

5 For items linked to an app on your MacBook, click on an item to view the information in the relevant app

6 For items from the web, click on an item to view the related web page(s)

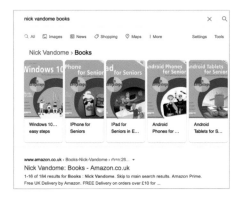

Click on a folder in Step 7 to open it directly in the Finder.

7 For other locations within your MacBook, click on an item to see the details in the main window

8 For items found in the Mail app, click on an email to view its details in the main window

Finder search

This is the Search box in the top right-hand corner of the Finder and can be used to search for items within it. See page 94 for details about using Finder search.

Using Siri

Siri is the digital voice assistant that can be used to vocally search for a wide range of items from your MacBook and the web. Also, the results can be managed in innovative ways so that keeping up-to-date is easier than ever.

Setting up Siri
To set up and start using Siri:

Hot tip

Check **On** the **Show Siri in menu bar** box to show the Siri icon in the top right-hand corner of the main Apple Menu bar.

☑ Show Siri in menu bar

1 Open **System Preferences** and click on the **Siri** button

2 Check **On** the **Enable Ask Siri** box

3 Make selections here for how Siri operates, including language and the voice used by Siri

4 Click on the **Siri** icon on the Dock or in the Menu bar to open Siri

Hot tip

An internal or external microphone is required in order to ask Siri questions.

5 The Siri window opens, ready for use

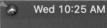

Searching with Siri

Siri can respond to most requests, including playing a music track, checking the weather, opening photos, and adding calendar events. If you open Siri but do not ask a question, it will prompt you with a list of possible queries. Click on the microphone icon at the bottom of the panel to ask a question. It is also possible to ask questions of Siri while you are working on another document.

Siri can search for a vast range of items, from your MacBook or the web. Some useful Siri functions include:

Searching for documents
This can be refined by specifying documents covering a specific date or subject:

1 Ask Siri to find and display documents with certain criteria; e.g. created by a specific person

2 Ask Siri to refine the criteria; e.g. only include documents created in a specific app. Click on an item to open it in its related app

It is possible to ask Siri to search for specific types of documents; e.g. those created in Pages, or PDF files.

...cont'd

Drag and drop results

Once search results have been displayed by Siri they can be managed in a variety of ways. One of these is to copy an item in the search results into another app, by dragging and dropping:

1 Ask Siri to display a certain type of item, such as a map location or an image

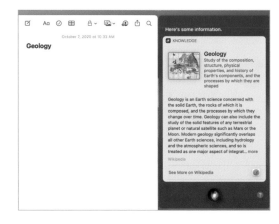

2 Drag the resulting item into another app, such as a word processing app or the Notes app

Multitasking

The Siri search window can be left open on the Desktop so that you can ask Siri a question while you are working on something else; e.g. while you are working on a document, ask Siri to display a relevant web page.

Accessibility

In all areas of computing it is important to give as many people as possible access to the system. This includes users with visual impairments and also people who have problems using the mouse and keyboard. In macOS this is achieved through the functions of the **Accessibility** system preferences. To use these:

1 Click on the **Accessibility** button in **System Preferences**

2 Click on the **Display** button for options for changing the display colors and contrast, and for increasing the cursor size

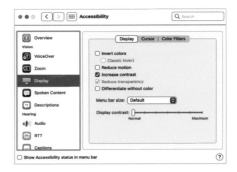

3 Click on the **Zoom** button for options to zoom in on the screen

4 Click on the **VoiceOver** button to enable VoiceOver, which provides a spoken description of what is on the screen

Don't forget

Experiment with the VoiceOver function, if only to see how it operates. This will give you a better idea of how visually-impaired users access information on a computer.

...cont'd

5 Click on the **Audio** button to select an on-screen flash for alerts, and how sound is played

Don't forget

The **Audio** accessibility option has a link to additional options within its own system preferences.

6 Click on the **Keyboard** button to access options for customizing the keyboard

7 Click on the **Pointer Control** button to access options for customizing the mouse and trackpad

8 Click on the **Voice Control** button to select options for using spoken commands

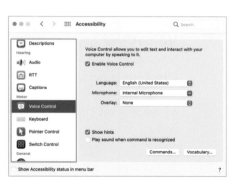

The Spoken Word

macOS Big Sur not only has numerous apps for adding text to documents, emails and messages; it also has a Dictation function so that you can speak what you want to appear on screen. To set up and use the Dictation feature:

1 Click on the **Dictation** tab in **System Preferences** > **Keyboard**

2 Click on the **On** button to enable Dictation

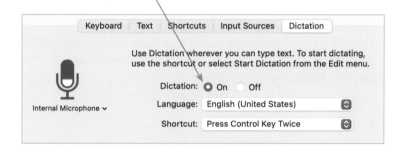

3 Click on the **Enable Dictation** button

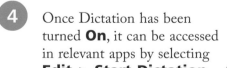

Hot tip

Punctuation can be added with the Dictation function, by speaking commands such as "comma" or "question mark". These will then be converted into the appropriate symbols.

51

4 Once Dictation has been turned **On**, it can be accessed in relevant apps by selecting **Edit** > **Start Dictation...** from the Menu bar

Start Dictation...

5 Start talking when the microphone icon appears. Click **Done** when you have finished recording your text

Shutting Down

The Apple menu (which can be accessed by clicking on the Apple icon at the top-left corner of the Desktop, or any subsequent macOS window) has been standardized in macOS. This means that it has the same options regardless of the app in which you are working. This has a number of advantages, not least being the fact that it makes it easier to shut down your MacBook. When shutting down, there are four options that can be selected:

When shutting down, make sure you have saved all of your open documents, although Big Sur will prompt you to do this if you have forgotten.

- **Sleep**. This puts the MacBook into hibernation mode; i.e. the screen goes blank and the hard drive becomes inactive. This state is maintained until the mouse is moved or a key is pressed on the keyboard. This then wakes up the MacBook and it is ready to continue work. It is a good idea to add a login password for accessing the MacBook when it wakes up, otherwise other people could wake it up and gain access to it. To add a login password, go to **System Preferences > Security & Privacy** and click on the **General** tab. Check **On** the **Require password** checkbox and select from one of the timescale options (**immediately** is best).

- **Restart**. This closes down the MacBook and then restarts it again. This can be useful if you have added new software and your computer requires a restart to make it active.

- **Shut Down**. This closes down the MacBook completely once you have finished working.

macOS Big Sur has a **Resume** feature whereby your MacBook opens up in the same state as when you shut it down. See page 80 for details.

- **Log Out**. This logs you out of your current session and closes down your open apps. You can then log back in without turning off your MacBook and return to your previously-open apps by using the **Resume** function; see tip.

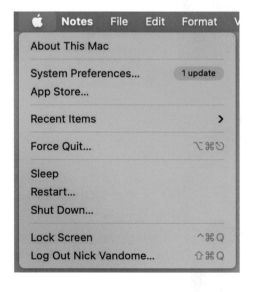

4 Getting Up and Running

This chapter looks at some of the essential features of Big Sur. These include the Dock for organizing and accessing all of the elements of your MacBook, and items for arranging folders and files. It also covers Apple's online sharing service, iCloud, which can be used to back up a variety of content and also share items such as music, books, calendars and apps with other family members.

The Dock has been updated in macOS Big Sur, to standardize the appearance of macOS apps.

The Dock is always displayed as a line of icons, but this can be orientated either vertically or horizontally (see next page).

Items on the Dock can be opened by clicking on them once, rather than having to double-click on them. Once they have been selected, the icon bobs up and down until the item is available.

Introducing the Dock

The Dock is one of the main organizational elements of macOS. Its main function is to help organize and access apps, folders and files. In addition, with its rich translucent colors and elegant graphical icons, it also makes for an aesthetically-pleasing addition to the Desktop. The main things to remember about the Dock are:

● It is divided into two: apps go on the left of the dividing line; all other items go on the right.

● It can be customized in a number of ways.

By default, the Dock appears at the bottom of the screen:

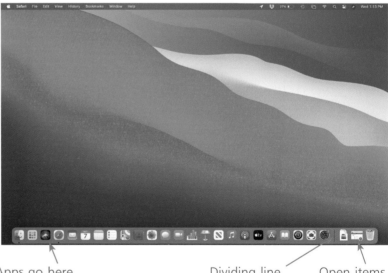

Apps go here   Dividing line   Open items

If an app window is closed, the app remains open and the window is placed within the app icon on the Dock. If an item is minimized, it goes on the right of the Dock dividing line.

Setting Dock Preferences

As with most elements of macOS, the Dock can be modified in several ways. This can affect both the appearance of the Dock and the way it operates. To set Dock preferences:

1 Select **System Preferences > Dock & Menu Bar**

Dock & Menu Bar

2 The Dock preferences allow you to change its size, orientation, the way icons appear with magnification, and effects for when items are minimized

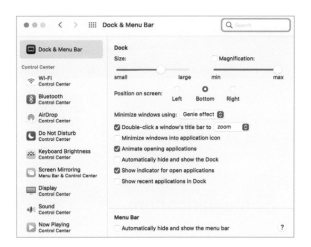

For instance, the size of the Dock can be reduced so that it only takes up a proportion of the width of the screen:

Don't forget

You will not be able to make the Dock size so large that some of the icons would not be visible on the Desktop. By default, the Dock is resized so that everything is always visible.

...cont'd

The **Position on screen** options enable you to place the Dock on the left, right or bottom of the screen:

Beware

The Dock cannot be moved by dragging it physically; this can only be done in the Dock preferences window.

1. Drag the **Dock Size** slider to increase or decrease the size of the Dock

Don't forget

When the cursor is moved over an item in the Dock, the name of that item is displayed above it.

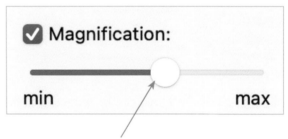

2. Check **On** the **Magnification** box and drag the slider to determine the size to which icons are enlarged when the cursor is moved over them

The effects that are applied to items when they are minimized is one of the features of macOS (it is not absolutely necessary but it sums up the Apple ethos of trying to enhance the user experience as much as possible).

The **Genie effect** shrinks the item to be minimized like a genie going back into its lamp:

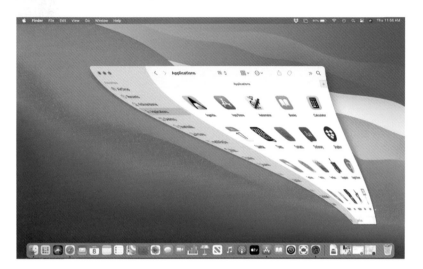

Hot tip

Open windows can also be minimized by double-clicking on their title bar (the thin bar at the top of the window, next to the three window buttons).

Manual resizing

In addition to changing the size of the Dock by using the Dock preferences dialog box, it can also be resized manually:

1 Drag vertically on the Dock dividing line to increase or decrease its size

Stacks on the Dock

Stacking items

To save space, it is possible to add folders to the Dock, from where their contents can be accessed. This is known as Stacks. By default, a Stack for downloaded files is created on the Dock. To use Stacks:

1 To create a new Stack, drag a folder onto the Dock. Stacked items are placed on the right of the Dock dividing line

Stacks can be removed from the Dock by dragging them over the Desktop, until the **Remove** button appears. At this point, release the Stack icon.

2 Click on a Stack to view its contents

3 Stacks can be viewed as:

● A grid.

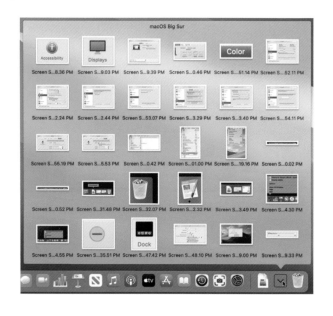

- A fan, depending on the number of items it contains.

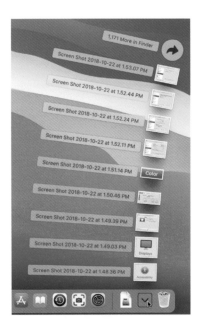

Move the cursor over a Stack on the Dock and press **Ctrl** + **click** to access options for how that Stack is displayed.

- A list. Click on a folder to view its contents within a Stack, then click on files to open them in their relevant app.

Any new items that are added to the folder will also be visible through the Stack.

Dock Menus

One of the features of the Dock is that it can display contextual menus for selected items. This means that it shows menus with options that are applicable to the item being accessed. This can only be done when an item has been opened.

Click on **Quit** on the Dock's contextual menu to close an open app or file, depending on which side of the dividing bar the item is located.

1 Click and hold on the black dot below an app's icon to display an item's individual context menu

2 Click on **Options**, then **Show in Finder** to see where the item is located on your computer

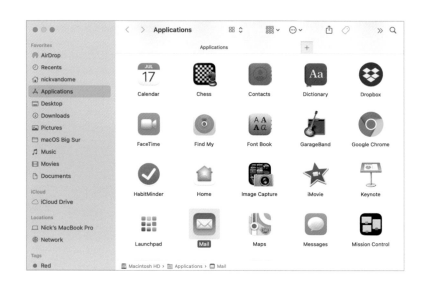

Working with Dock Items

Adding items

As many items as you like can be added to the Dock; the only restriction is the size of monitor in which to display all of the Dock items (the size of the Dock can be reduced to accommodate more icons, but you have to be careful that all of the icons are still legible). To add items to the Dock:

Icons on the Dock are shortcuts to the related item rather than the item itself, which remains in its original location.

1 Locate the required item in the Finder and drag it onto the Dock. All of the other icons move along to make space for the new one

Keep in Dock

Every time you open a new app, its icon will appear in the Dock for the duration that the app is open, even if it has not previously been put in the Dock. If you then decide that you would like to keep it in the Dock, you can do so as follows:

1 Click and hold on the dot below an open app

2 Click on **Options**, then **Keep in Dock** to ensure the app remains in the Dock when it is closed

...cont'd

Removing items

Any item, except the Finder, can be removed from the Dock. However, this does not remove it from your computer; it just removes the shortcut for accessing it. You will still be able to locate it in its folder in the Finder and, if required, drag it back onto the Dock. To remove items from the Dock:

Hot tip

When an icon is dragged from the Dock, it has to be moved a reasonable distance before the **Remove** alert appears.

1 Drag the item away from the Dock until it displays the **Remove** tag. The item disappears once the cursor is released. All of the other icons then move up to fill in the space

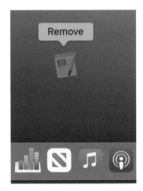

Removing open apps

You can remove an app from the Dock, even if it is open and running. To do this:

1 Drag an app off the Dock while it is running. Initially, the icon will remain on the Dock because the app is still open

Don't forget

If **Keep in Dock** has been selected for an item (see page 61) the app will remain in the Dock even when it has been closed.

2 When the app is closed (click and hold on the dot underneath the item and select **Quit**) its icon will be removed from the Dock

Trash

The Trash folder is a location for placing items that you do not need anymore. However, when items are placed in the Trash, they are not removed from your computer. This requires another command, as the Trash is really a holding area before you decide you want to remove items permanently. The Trash can also be used for ejecting removable disks attached to your MacBook.

Sending items to the Trash
Items can be sent to the Trash by dragging them from the location in which they are stored:

Items can also be sent to the Trash by selecting them and then selecting **File** > **Move to Trash** from the Menu bar.

1 Drag an item over the **Trash** icon to place it in the Trash folder

Screen Shot 2020-10...8.06 AM

2 Click once on the **Trash** icon on the Dock to view its contents

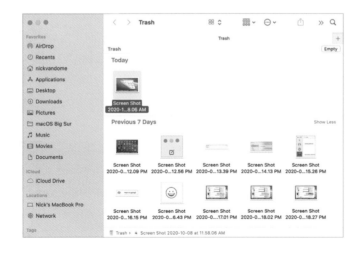

All of the items within the Trash can be removed in a single command: select **Finder** > **Empty Trash** from the Menu bar to remove all of the items in the Trash folder.

Control Center

The Control Center has been a feature of Apple's mobile devices – the iPhone and iPad – for a number of years, and it is now available within macOS Big Sur. It is a panel containing shortcuts to some of the most commonly-used options within **System Preferences**.

The inclusion of the Control Center is a new feature in macOS Big Sur.

Do Not Disturb can be used to specify times during which notifications and FaceTime calls can be muted. Screen Mirroring can be used to display the MacBook's screen on another screen.

AirDrop is the functionality for sharing items wirelessly between compatible devices. Click once on the **AirDrop** button in the Control Center and specify whether you want to share with **Contacts Only** or **Everyone**. Once AirDrop is set up, you can use the **Share** button in compatible apps to share items such as photos with any other AirDrop users in the vicinity.

Accessing the Control Center
The Control Center can be accessed from the top toolbar on the Desktop or from within any app. To do this:

1 To open the Control Center, click on this button on the top toolbar

2 Click on these buttons to access options for **Do Not Disturb**; **Keyboard Brightness**; and **Screen Mirroring**

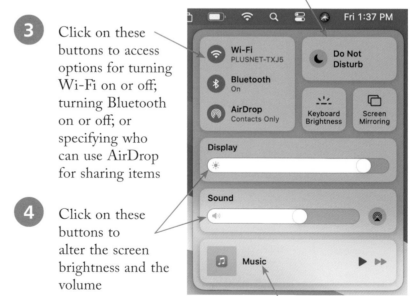

3 Click on these buttons to access options for turning Wi-Fi on or off; turning Bluetooth on or off; or specifying who can use AirDrop for sharing items

4 Click on these buttons to alter the screen brightness and the volume

5 Click on this button to access the Music widget for playing and controlling tracks from the Music app

...cont'd

Managing the Control Center

The items within the Control Center can be managed within System Preferences:

1 Open System Preferences and click on the **Dock & Menu Bar** button

2 The items in the Control Center are listed in the left-hand sidebar

3 Click on an item to view its options. If it is an item that is always in the Control Center, the only option will be to also **Show in Menu Bar**

Wi-Fi
Wi-Fi is always available in Control Center.

☐ Show in Menu Bar

4 Swipe up in the left-hand sidebar to see additional options that can be added to the Control Center, under the **Other Modules** heading

Other Modules

🧑‍🦽 Accessibility Shortcuts

🔋 Battery

👤 Fast User Switching

5 Click on one of the Other Modules options and check **On** the **Show in Control Center** checkbox to add the item to the Control Center

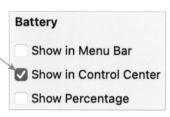

Battery

☐ Show in Menu Bar

☑ Show in Control Center

☐ Show Percentage

Don't forget

The items in the left-hand sidebar under the **Control Center** heading are always available in the Control Center.

Hot tip

For the Battery option, check **On** the **Show Percentage** checkbox to show the amount of battery power left for a MacBook as a percentage, when it is using battery power rather than mains power.

The Notification Center has been updated in macOS Big Sur.

Notifications can be accessed regardless of the app in which you are working, and they can be actioned directly without having to leave the active app.

Twitter and Facebook feeds can also be displayed in the Notification Center, if you have accounts with these sites.

Notifications and Widgets

The Notification Center provides a single location to view all of your emails, messages, updates and alerts. Notifications appear at the top right-hand corner of the screen. In macOS Big Sur, the Notification Center now includes widgets that can display a range of useful information. The widgets can be edited within the Notification Center itself, and the apps that display notifications are set up within System Preferences. To do this:

1 Open **System Preferences** and click on the **Notifications** button

2 The items that will appear in the Notification Center are listed here. Click on an item to select it and set its notification options

3 To disable an item so that it does not appear in the Notification Center, select it as above and check **Off** the **Show in Notification Center** box

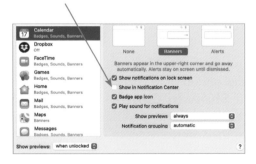

Viewing notifications

Notifications appear in the Notification Center. The way they appear can be determined in System Preferences:

1 Select an alert style. A banner alert comes up on the screen and then disappears after a few seconds

2 The **Alerts** option shows the notification, and it stays on screen

until dismissed (such as this one for Reminders)

3 Click on the day and time icon in the top right-hand corner of the Menu bar to view all of the items in the Notification Center. Click on it again to hide the Notification Center

4 Notifications that have been set up in System Preferences appear at the top of the Notification Center. The widgets appear below any notifications (see pages 68-69 for details)

The Notification Center can also be displayed using a trackpad or Magic Trackpad by dragging with three fingers from right to left, starting from the far-right edge.

Software updates can also appear in the Notification Center, when they are available.

67

Widgets in the Notification Center is a new feature in macOS Big Sur.

Click on this button in the top left-hand corner of a widget in Step 2 to remove it from the Notification Center:

...cont'd

Using widgets

The widgets that appear in the Notification Center display can be managed and edited directly from the Notification Center panel. To do this:

1 Swipe down to the bottom of the Notification Center in Step 4 on page 67 and tap on the **Edit Widgets** button

2 The existing widgets in the Notification Center are shown in the right-hand panel. The middle panel displays widgets that can be added to the Notification Center

3 All of the available widgets are shown in the left-hand panel. Click on one to view details about it in the middle panel, so that it can be added to the Notification Center, if required

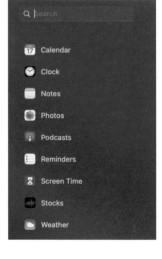

4 Click on a widget in the middle panel to add it to the Notification Center, or move the cursor over the widget and tap on the green **+** button

5 Some of the widgets have an option for adding them to the Notification Center at different sizes. If this is available for a widget, these buttons will appear below the widget in the middle panel in Step 2 on the previous page. The options are for **Small**, **Medium** or **Large**

Hot tip

Several versions of the same widget can be added to the Notification Center, at different sizes, if required.

Hot tip

Some widgets can be edited to display different information. For instance, the Weather widget can display forecasts from numerous locations around the world. To see if a widget is editable, move the cursor over it in the right-hand panel in Step 2 on the previous page. If the widget is editable, click on the **Edit Widget** button.

About iCloud

Cloud computing is an attractive proposition and one that has gained great popularity in recent years. As a concept, it consists of storing your content on an external computer server. This not only gives you added security in terms of backing up your information; it also means that the content can then be shared over a variety of devices.

iCloud is Apple's consumer cloud computing product that consists of online services such as email; a calendar; notes; contacts; and saving photos and documents. iCloud provides users with a way to save and back up their files and content to the online service, and then use them across their Apple devices such as other Mac computers, iPhones, iPads and iPod Touches.

About iCloud

Once iCloud has been set up (see next page) it can be accessed from within System Preferences:

1 Click on the **Apple ID** button

2 Click on the **iCloud** button

You can use iCloud to save and share the following between your different devices, with an Apple ID account (see next page):

- Photos
- Mail and Safari
- Documents
- Backups
- Notes
- Reminders
- Contacts and Calendar

When you save an item to iCloud it automatically pushes it to all of your other compatible devices; you do not have to manually sync anything as iCloud does it all for you.

Don't forget

The standard iCloud service is free, and this includes an iCloud email address and 5GB of online storage. (*Correct at the time of printing.*)

Don't forget

There is also a version of iCloud for Windows, which can be accessed for download from the Apple website at https://support.apple.com/en-us/HT204283

Setting up iCloud

To use iCloud with macOS Big Sur you need to first have an Apple ID. This is a service you can register for to be able to access a range of Apple facilities, including iCloud. You can register with an email address and a password. When you first start using iCloud you will be prompted for your Apple ID details. If you do not have an Apple ID you can create one at this point.

1 Open System Preferences and click on the **Sign In** button next to **Sign in to your Apple ID**

When you have an Apple ID and an iCloud account, you can also use the iCloud website to access your content. Access the website at **www.icloud.com** and log in with your Apple ID details.

2 Sign in with your Apple ID, or

3 Click on the **Create Apple ID...** link and follow the steps to create your Apple ID

4 Open System Preferences and click on the **iCloud** button

5 Check **On** the items you want included within iCloud. All of these items will be backed up and shared across all of your Apple devices

The online iCloud service includes your online email service; Contacts; Calendars; Reminders; Notes; and versions of Pages, Keynote and Numbers. You can log in to your iCloud account from any internet-enabled device.

71

About iCloud Drive

One of the options in the iCloud section is for iCloud Drive. This can be used to store documents and other content so that you can use them on any other Apple devices that you have, such as an iPhone or an iPad. With iCloud Drive you can start work on a document on one device, and continue on another device from where you left off. To set up iCloud Drive:

Pages is the Apple app for word processing; Numbers for spreadsheets; and Keynote for presentations. These can all be downloaded from the App Store, if they do not come pre-installed on your MacBook.

1 Click on the **Apple ID** > **iCloud** button in System Preferences, as shown on page 70

2 Check On **iCloud Drive** to activate iCloud Drive, for use with compatible apps, such as Pages, Numbers and Keynote

Using iCloud Drive
To work with files in iCloud Drive, once it has been set up:

If the **Desktop & Documents Folders** option is selected for iCloud Drive this can take up a lot of iCloud storage, as all items in these locations will be saved to iCloud Drive, which counts toward your iCloud storage limit.

1 In the Finder sidebar, click on the **iCloud Drive** button

2 Certain iCloud Drive folders are already created, based on the apps that you have selected on page 71. These are the default folders into which content from their respective apps will be placed (although others can also be selected, if required). Double-click on a folder to view its contents

3 Double-click on one of the folders in the previous step to view its details. Double-click on an item to open it in its related app

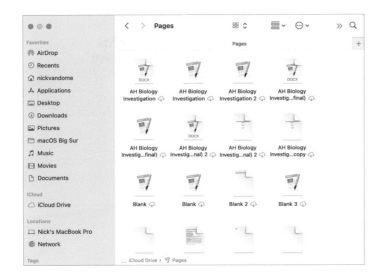

Another useful iCloud function is the iCloud Keychain (**System Preferences** > **Apple ID** > **iCloud** (in the left-hand sidebar) and check **On** the **Keychain** option). If this is enabled, it can keep all of your passwords and credit card information up-to-date across multiple devices and remember them when you use them on websites. The information is encrypted and controlled through your Apple ID.

73

4 Use these three buttons on the top toolbar

to, from left to right: show the items in the window as icons, a list, in columns or as a gallery; select options for how the items in the window are grouped; and perform file management tasks, including creating new folders and organizing the items in the window

5 To save files into an iCloud Drive folder, select **File** > **Save**, or **Save As**,

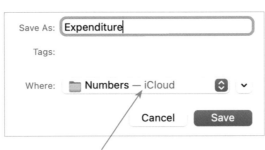

from the Menu bar, click on the **iCloud Drive** button in the Finder sidebar, and navigate to the required folder for the file

Continuity

One of the main themes of macOS Big Sur – and iOS/iPadOS for mobile devices – is to make all of your content available on all of your Apple devices. This is known as Continuity: when you create something on one device you can then pick it up and finish it on another. This is done through iCloud. To use this feature:

1 Ensure the app has iCloud activated, as on page 70

2 Create the content in the app on your MacBook

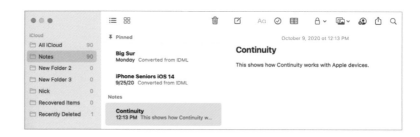

3 Open the same app on another Apple device; e.g. an iPad. The item created on your MacBook should be available to view and edit. Any changes will then show up on the file on your MacBook too

Hot tip

It is also possible to continue an email with the Continuity feature. First, create it on your MacBook and then close it. You will be prompted to save the email as a draft and, if you do this, you will be able to open it from the **Drafts** mailbox on another Apple device.

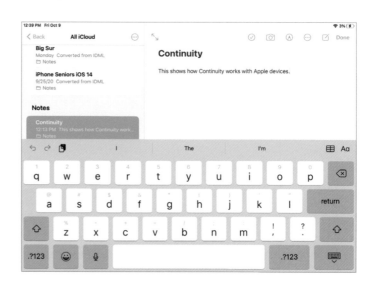

Handoff

Handoff is one of the key features of Continuity, and it displays icons of items that you have opened on another device, such as Safari web pages. Handoff does not work with all devices, and it only works if both devices have OS X Yosemite (or later) and iOS 8 (or later), for mobile devices.

To use Handoff you will need to do the following:

- Your MacBook must be running OS X Yosemite (or later) and your mobile device (iPhone 5 and later, iPad 4th generation and later, all models of iPad mini and the 5th-generation iPod Touch) must have iOS 8 (or later).

- Your MacBook has to support Bluetooth 4.0, which means that most pre-2012 Macs are not compatible with Handoff.

To check if your MacBook supports Handoff:

The apps that work with Handoff are Mail, Safari, Maps, Messages, Reminders, Calendar, Contacts, Notes, Pages, Numbers, and Keynote.

1 Select **Apple menu > About This Mac > System Report**. Click on **Bluetooth** to see if Handoff is supported

Bluetooth Low Energy Supported:	Yes
Handoff Supported:	Yes
Instant Hot Spot Supported:	Yes

2 Turn on Bluetooth on your MacBook (in **System Preferences**) and on your mobile device (in **Settings**)

3 Turn on Handoff on your MacBook (**System Preferences > General** and check On **Allow Handoff Between this Mac and your iCloud Devices**) and on your mobile device (**Settings > General > Handoff**)

If Handoff is not working, try turning both devices off and on, and do the same with Bluetooth. Also, try logging out of, and then back in to, your iCloud account on both devices.

4 When Handoff is activated, compatible apps will be displayed at the left-hand side of the Dock when they have been opened on another device

About Family Sharing

Being able to share digital content with other people is an important part of the digital world, and with macOS Big Sur and iCloud, the Family Sharing function enables you to share items that you have downloaded from the relevant online Apple store with your Apple ID, such as music and movies, with up to five other family members, as long as they have an Apple ID. To set up Family Sharing:

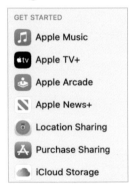

Don't forget

Once Family Sharing has been set up, the items that will be shared are listed, including subscriptions and new purchases, in the left-hand sidebar in Step 2. Click on an item to access it: once this has been done it will be available to other Family Sharing members.

Don't forget

A Family Sharing invitation can be sent by email or using the Messages app, or by using AirDrop with another compatible Apple device in close proximity.

1 Access the Apple ID section within System Preferences and click on the **Family Sharing** button

Family Sharing

2 Click on the **Get Started** button (the initial setup for Family Sharing involves a step-by-step wizard, during which you can include a payment method for Family Sharing items)

3 Click on the **Invite People** button

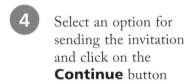

4 Select an option for sending the invitation and click on the **Continue** button

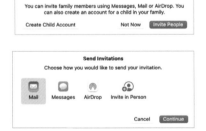

5 Select the person who you want to invite, add text as required, and send the invite. The recipient has to accept the invite to join your Family Sharing group

Using Family Sharing

Once you have set up Family Sharing and added family members, you can start sharing a selection of items.

Sharing photos

Photos can be shared with Family Sharing, with the Family album that is created automatically within the Photos app.

1 Click on the **Photos** app

2 Click on the **Shared Albums** button underneath the Albums heading in the left-hand sidebar

3 The **Family** album is already available in the **Shared Albums** section. Double-click on the Family album to open it

4 Click on the **Add photos and videos** option

Add photos and videos

Each family member receives a notification when a new photo is added to a family album.

5 Click on the photos you want to add, and click on the **Add** button

6 The photos are added to the Family album and all members of Family Sharing will be able to view them

More than one family member can use content in the Family Sharing group at the same time.

When a Family Sharing calendar event is created, other people in your Family Sharing circle will have this event added to their Family calendar, and they will also be sent a notification.

To change the color of a calendar, click on the **Calendars** button at the top left of the window. **Ctrl + click** on a calendar name and select a color from the bottom of the panel, or click on **Custom Color** to choose from the full spectrum.

...cont'd

Sharing apps, music, books and movies

Family Sharing means that all members of the group can share purchases from the iTunes Store, the App Store or the Books app. This is done from the Purchased section of each app:

1 Open the relevant app and access the Purchased section. (For the App Store,

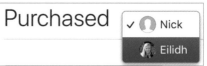

click on the Account icon and click on the Purchased button; for the iTunes Store, click on the Purchased button on the bottom toolbar; for the Books app, click on the Account icon)

2 For all three apps, click on a member under **Family Purchases** to view their purchases and download them, if required, by tapping once on this button

Sharing calendars

Family Sharing also generates a Family calendar that can be used by all Family Sharing members:

1 Open the **Calendar** app

2 Double-click on a date to create a **New Event**

3 Click here and click on the **Family** calendar

4 Complete the details for the event. It will be added to your calendar, with the Family color tag (see page 93)

Finding lost family devices

Family Sharing also makes it possible to see where everyone's devices are, which can be useful for locating people, but particularly so if a device belonging to a Family Sharing member is lost or stolen, including your own. To use this functionality:

1 Ensure that the **Find My Mac** function is turned on in the **iCloud** system preferences, and log in to your online iCloud account at **www.icloud.com**

2 Click on the **Find iPhone** button (this works for other Apple devices, too)

3 Devices that are turned on, online and with iCloud activated are shown by green dots

4 Click on a green dot to display information about the device. Click on the **i** symbol to access options for managing the device remotely

5 There are options to send an alert sound to the device, lock it remotely or erase its contents (if you are concerned about it having fallen into the wrong hands)

Hot tip

Lost or missing Apple devices can be searched for from a MacBook, using the **Find My** app. The missing devices have to be linked to the same Apple ID or be connected via Family Sharing.

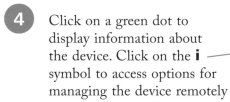

Resuming

One of the chores of computing is that when you close down your computer you have to first close down all of your open documents and apps, and then open them all again when you turn your machine back on again. However, macOS has a Resume feature that allows you to continue working exactly where you left off, even if you turn off your computer. To activate this:

1 Before you close down, all of your open documents and apps will be available

2 Select the **Shut Down...** or **Restart...** option from the Apple menu

3 Make sure this box is checked **On** (this will ensure that all of your items will appear as before, once the MacBook is closed down and then opened again)

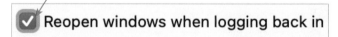

4 Confirm the **Shut Down** (or **Restart**) command

If the Shut Down (or Restart) button is not clicked on in Step 4, the action will be performed automatically after a period of one minute.

5 Finder

The main way of moving around Big Sur is with the Finder. This enables you to access items and organize your apps, folders and files. This chapter looks at how to use the Finder and get the most out of this powerful tool. It covers how to customize the interface, numerous options for working with folders and files, and also sharing items with other apps directly from the Finder.

Working with the Finder

If you were only able to use one item on the Dock, it would be the Finder. This is the gateway to all of the elements of your MacBook. It is possible to get to selected items through other routes, but the Finder is the only location where you can gain access to everything on your system. If you ever feel that you are getting lost within macOS, click on the Finder and then you should begin to feel more at home. To access the Finder:

1 Click once on this icon on the Dock

Overview

The Finder has its own toolbar; a sidebar from which items can be accessed; and a main window where the contents of selected items can be viewed:

Forward and back   View options   Actions button Search

View recent files

Folders are displayed here

Sidebar

Tags

Main window

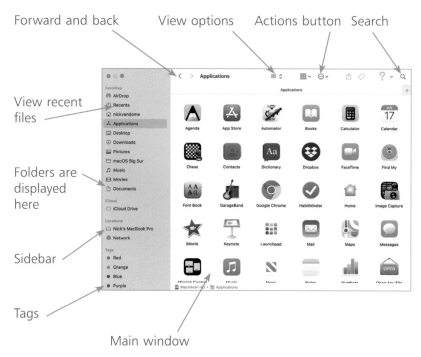

Don't forget

A link to iCloud Drive is also included in the Finder sidebar.

Don't forget

The **Actions** button has options for displaying information about a selected item, and also options for how it is displayed within the Finder. See page 98 for more details about using the Actions button.

Finder Folders

Recents

This contains all of the latest files on which you have been working. They are sorted into categories according to file type so that you can search through them quickly. This is an excellent way to locate items without having to look through a lot of folders. To access this:

1 Click on this link in the Finder sidebar to access the contents of your **Recents** folder

> 🕐 Recents

2 All of your recent files are displayed in individual categories. Click on the headings at the top of each category to sort items by those criteria

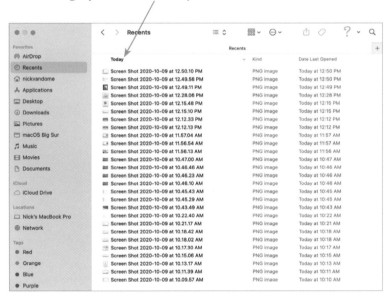

3 Double-click on an item to open it from the Finder

The Finder is always open (as denoted by the black-dot graphic underneath its icon on the Dock) and it cannot readily be closed down or removed.

The Finder sidebar has the macOS Big Sur transparency feature, so that you can see some of the open window or Desktop behind it.

To change the display of folders in the sidebar, click on the **Finder** menu on the top toolbar. Select **Preferences** and click on the **Sidebar** tab. Under **Show these items in the sidebar:**, select the items you want included.

...cont'd

Home folder

This contains the contents of your own Home directory, containing your personal folders and files. macOS inserts some pre-named folders that it thinks will be useful, but it is possible to rearrange or delete these as you please. It is also possible to add as many more folders as you want.

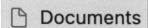

Applications

This folder contains all of the applications on your MacBook. They can also be accessed from the Launchpad, as shown on pages 118-119.

Documents

This is part of your Home folder but is put on the Finder sidebar for ease of access. New folders can be created for different types of documents.

1 Click on your Apple ID account name in the sidebar to access the contents of your **Home** folder

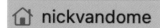
nickvandome

2 The Home folder contains the **Public** folder, which can be used to share files with other users if the computer is part of a network

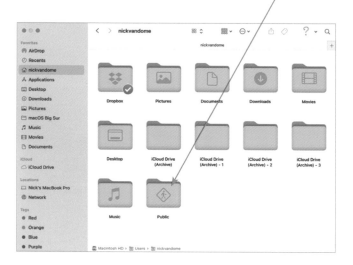

When you are creating documents, macOS by default recognizes their type, and then when you save them suggests the most applicable folder in which to save them.

3 To add a new folder, click the **Actions** button (see page 98), then **New Folder** (or **Ctrl + click** in the window and select **New Folder**). The new folder will open in the main window, named "untitled folder". Overtype the name with one of your choice

4 To delete a folder from the Finder sidebar, **Ctrl + click** on it and select **Remove from Sidebar**

Finder Views

The way in which items are displayed within the Finder can be amended in a variety of ways, depending on how you want to view the contents of a folder. Different folders can have their own viewing options applied to them, and these will stay in place until a new option is specified.

Back button
When working within the Finder, each new window replaces the previous one, unless you open a new app. This prevents the screen becoming cluttered with dozens of open windows, as you look through various Finder windows for a particular item. To ensure that you never feel lost within the Finder structure, there is a Back button on the Finder toolbar that enables you to retrace the steps you have taken.

Hot tip

Select an item within the Finder window and click on the space bar to view its details.

1 Navigate to a folder within the Finder (in this example, the **macOS Big Sur** folder contained within **Pictures**)

2 Click on the **Back** button to move back to the previously-visited window (in this example, the main **Pictures** window)

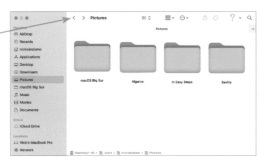

Beware

If you have not opened any Finder windows, the Back button will not operate.

...cont'd

View options

It is possible to customize some of the options for how items are viewed in the Finder. To do this:

Hot tip

The **Group By** option in Step 2 can be used to arrange icons into a specific group (e.g. by name or type) or to snap them to an invisible grid so that they have an ordered appearance.

1 Click on this button on the Finder toolbar

2 Select **Show View Options** to access the options for customizing the Finder view

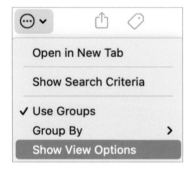

3 Drag this slider to set the icon size

Hot tip

A very large icon size can be useful for people with poor eyesight, but it takes up a lot more space in a window.

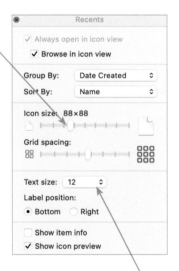

4 Select options here for the way items are arranged and displayed in Finder windows

Finder views

There are several options for displaying folders and files within the Finder. To access them:

1 Click on this button on the Finder toolbar

2 Select one of the view options, from **as Icons**, **as List**, **as Columns** and **as Gallery**

✓ as Icons
as List
as Columns
as Gallery

The button in Step 1 changes in appearance, depending on the selection in Step 2.

Icon view

List view

Columns view

If a folder has a right-pointing arrow next to it, this indicates that there is content in the folder. Click on the arrow to view the contents of the folder.

> 📄 **Documents**

87

Gallery View

Another of the view options on the Finder toolbar is the Gallery view: this is a view that has considerable functionality in terms of viewing details of specific files and also applying some quick editing functions to them without having to open the files. To use the Gallery view:

1 Select a file in either Icon, List or Column view and click on the **as Gallery** option from the button on the top of the toolbar, as shown on page 87

2 The item is displayed in the main window, with thumbnails of the other available files below it

3 The item that is displayed will change depending on the type of file that is selected

Hot tip

Gallery view is a good way to view photos on your MacBook, without having to open the Photos app.

Quick Look

Through a Finder option called Quick Look, it is possible to view the content of a file without having to first open it. To do this:

1 Select a file in any of the Finder views

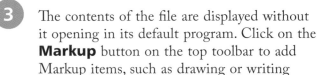

2 Press the space bar on the keyboard

3 The contents of the file are displayed without it opening in its default program. Click on the **Markup** button on the top toolbar to add Markup items, such as drawing or writing

4 Click on the cross to close Quick Look, or view the item in full screen

Hot tip

The Markup button provides a range of options for drawing or writing on an item, including types of pens, shapes, borders, colors and text styles.

Hot tip

In Quick Look it is even possible to preview videos or presentations without having to first open them in their default app. If videos are being viewed, there will be a trim button so that the length of the video can be trimmed without having to open it in a specific app.

Finder Toolbar

Customizing the toolbar

As with most elements of macOS, it is possible to customize the Finder toolbar:

1 Select **View** > **Customize Toolbar...** from the Menu bar

Do not put too many items on the Finder toolbar, because you may not be able to see them all in the Finder window. If there are additional toolbar items, there will be a directional arrow indicating this. Click on the arrow to view the available items.

2 Drag items from the window into the toolbar

3 Alternatively, drag the default set of icons into the toolbar

4 Click **Done** at the bottom of the window

Finder Sidebar

Using the sidebar

The sidebar is the left-hand panel of the Finder that can be used to access items on your MacBook. To use this:

1 Click on a button on the sidebar

2 Its contents are displayed in the main Finder window

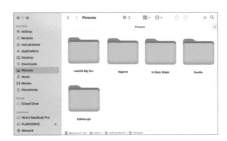

When you click on an item in the sidebar, its contents are shown in the main Finder window to the right.

Adding to the sidebar

Items that you access most frequently can be added to the sidebar.

1 Drag an item from the main Finder window onto the sidebar

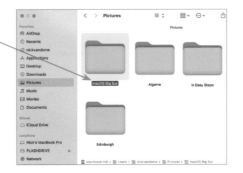

When items are added to the Finder sidebar, a shortcut – or alias – is inserted into the sidebar, not the actual item.

2 A blue line appears determining the location, and the item is added to the sidebar at this point. You can do this with apps, folders and files

Items can be removed from the sidebar by **Ctrl + clicking** on them and selecting **Remove from Sidebar** from the contextual menu.

Finder Tabs

Tabs in web browsers are now well established, whereby you can have several pages open within the same browser window. This technology is utilized in the Finder in macOS Big Sur with the use of Finder tabs. This enables different folders to be open in different tabs within the Finder, so that you can organize your content exactly how you want. To use tabs:

Hot tip

In macOS Big Sur, tabs are available in a range of apps that open multiple windows, including the Apple productivity apps: Pages, Numbers and Keynote. If the tabs are not showing, select **View** > **Show Tab Bar** from the app's Menu bar.

Don't forget

To specify an option for what appears as the default for a new Finder window, click on the **Finder** menu and click on **Preferences**, then the **General** tab. Under **New Finder windows show:**, select the default window to be used.

1 Select **View** > **Show Tab Bar** from the Finder Menu bar

2 Click on this button to add a new tab

3 At this point the content in the new tab is displayed for the window (see Don't forget tip)

4 Each tab view can be customized, and this is independent of the other tabs

Finder Tags

When creating content in macOS Big Sur you may find that you have documents of different types that cover the same topic. For instance, you may have work-related documents in Pages for reports, Keynote for presentations and Numbers for spreadsheets. With the Finder Tags function it is possible to link related content items through the use of colored tags. These can be added to items in the Finder and also in apps when content is created.

Hot tip

Tags can also be added to items by **Ctrl + clicking** on them and selecting the required tag from the menu that appears. Alternatively, they can be added from this button on the main Finder toolbar:

1. The tags are listed in the Finder sidebar. (If you can't see the list, hover the cursor over the Tags heading and then click on the **Show** button)

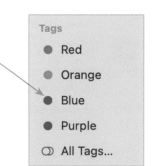

2. To give tags specific names, **Ctrl + click** on one and click on the **Rename** link

3. To add tags, select the required items in the Finder window

4. Drag the selected items over the appropriate tag

5. The tags are added to the selected items

Hot tip

Tags can also be added when documents are created in certain apps, such as Pages, Keynote and Numbers: select **File** > **Save**, click in the **Tags** box and select the required tag. Click on the **Save** button.

Finder Search

Searching electronic data is now a massive industry, with companies such as Google leading the way with online searching. On Macs it is also possible to search your folders and files, using the built-in search facilities: the Finder search; Siri (see pages 46-48); or Spotlight search (see pages 44-45).

Using Finder

To search for items within the Finder:

Don't forget

Try to make your search keywords and phrases as accurate as possible. This will create a better list of search results.

1 In the Finder window, enter the search keyword(s) in this box. Search options are listed below the keyword

2 Select an option for which category you want to search over

3 The search results are shown in the Finder window. Click on one of these buttons to search specific areas

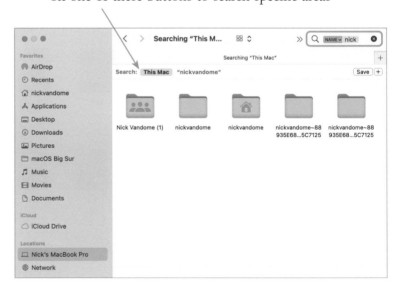

Don't forget

Both folders and files will be displayed in the Finder as part of the search results.

4 Double-click on an item to open it

Copying and Moving Items

Items can be copied and moved within macOS by using the copy and paste method, or by dragging.

Copy and paste

1 Select an item (or items) and select **Edit > Copy** from the Menu bar

Edit	View	Go	Window
Undo			⌘ Z
Redo			⇧ ⌘ Z
Cut			⌘ X
Copy 4 Items			⌘ C
Paste			⌘ V
Select All			⌘ A

2 Move to the target location and select **Edit > Paste Item(s)** from the Menu bar. The item is then pasted into the new location

Edit	View	Go	Window
Undo			⌘ Z
Redo			⇧ ⌘ Z
Cut			⌘ X
Copy			⌘ C
Paste 4 Items			⌘ V
Select All			⌘ A

3 To view the items that have been copied onto the Clipboard, select **Edit > Show Clipboard** from the Menu bar. The items are displayed in the Clipboard window

Edit	View	Go	Window
Undo			⌘ Z
Redo			⇧ ⌘ Z
Cut			⌘ X
Copy			⌘ C
Paste 4 Items			⌘ V
Select All			⌘ A
Show Clipboard			

```
● ● ●                    Clipboard
Screen Shot 2020-10-11 at 1.23.46 PM
Screen Shot 2020-10-11 at 1.24.32 PM
Screen Shot 2020-10-11 at 1.24.58 PM
Screen Shot 2020-10-11 at 1.25.41 PM
```

Dragging

Drag a file from one location to another to move it to that location. (This requires two or more Finder windows to be open, or drag the item over a folder on the sidebar.)

Don't forget

When an item is copied, it is placed on the Clipboard and remains there until another item is copied.

Hot tip

macOS Big Sur supports the Universal Clipboard, whereby items can be copied on a device such as an iPhone and then pasted directly into a document on a MacBook running macOS Big Sur. The process is the same as regular copy and paste, except that each operation is performed on a separate device and each app used has to be set up for iCloud. Also, Bluetooth and Wi-Fi have to be activated on both devices.

Working with Folders

When macOS Big Sur is installed, there are various folders that have already been created to hold apps and files. Some of these are essential (i.e. those containing apps), while others are created as an aid for where you might want to store the files that you create (such as the Pictures and Movies folders). Once you start working with macOS Big Sur you will probably want to create your own folders in which to store and organize your documents. This can be done on the Desktop or within any level of your existing folder structure. To create a new folder:

Don't forget

Folders are always denoted by a folder icon. This is the same regardless of the Finder view that is selected. The only difference is that the icon is larger in Icon view than in List or Column views.

Beware

You can create as many "nested" folders (i.e. folders within other folders) as you want. However, this makes your folder structure more complicated and, after a time, you may forget where all your folders are and what they contain.

Don't forget

Content can be added to an empty folder by dragging it from another folder and dropping it into the new one.

1 Access the location in which you want to create the new folder (e.g. your Home folder) and select **File** > **New Folder** from the Menu bar

2 A new, empty folder is inserted at the selected location (named "untitled folder")

3 Overtype the file name with a new one. Press **Enter** on the keyboard

4 Double-click on a folder to view its contents (at this point it should be empty)

Selecting Items

Apps and files within macOS folders can be selected by a variety of different methods.

Selecting by dragging

Click and drag the cursor to encompass the items to be selected. The selected items will become highlighted.

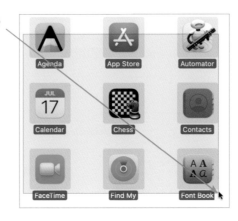

Once items have been selected, a single command can be applied to all of them. For instance, you can copy a group of items by selecting them and then applying the **Copy** command from the Menu bar.

Selecting by clicking

Click once on an item to select it, hold down **Shift**, and then click on another item in a list to select a consecutive group of items.

To select all of the items in a folder, select **Edit** > **Select All** from the Menu bar. The Select All command selects all of the elements within the active item. For instance, if the active item is a word processing document, the Select All command will select all of the items within the document; if it is a folder, it will select all of the items within that folder. You can also press **Command** + **A** on the keyboard to select all items.

To select a non-consecutive group, select the first item by clicking on it once, then hold down the Command key (**cmd ⌘**) and select the other required items. The selected items will appear highlighted.

Actions Button

The Finder Actions button provides a variety of options for any item, or items, selected in the Finder. To use this:

The icons on the Finder toolbar can be changed by customizing them – see page 90.

1 Select an item, or group of items, about which you want to find out additional information

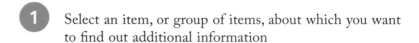

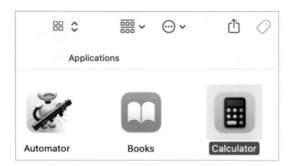

The Actions button can also be used for labeling items with Finder Tags. To do this, select the required items in the Finder and click one of the colored dots at the bottom of the Actions button menu. The selected tag will be applied to the item names in the Finder.

2 Click on the **Actions** button on the Finder toolbar

3 The available options for the selected item, or items, are displayed. These include **Get Info**, which displays additional information about an item such as file type, file size, creation and modification dates, and the default app for opening the item(s)

Sharing from the Finder

Also on the Finder toolbar is the Share button. This can be used
to share a selected item, or items, in a variety of ways appropriate
to the type of file that has been selected. For instance, a photo will
have options including social media sites such as Twitter, while
a text document will have fewer options. To share items directly
from the Finder:

1 Locate and select the item(s) that
you want to share

This button on the
Finder is for changing
the arrangement of
items within the Finder.
Click on this to access
arrangement options
such as Name, Date
and Size.

2 Click on the **Share** button on the
Finder toolbar and select one of the
options

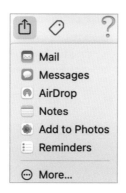

3 Click on the
More... button
in the previous
step to specify
more options for
sharing items.
Click on the
Share Menu
button in the left-
hand sidebar and
check **On** apps

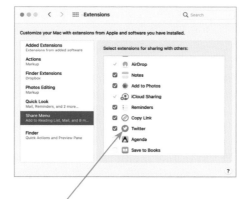

with which you want to share selected items, including
social media sites, if the relevant apps are installed

Menus

The main Apple Menu bar in macOS Big Sur contains a variety of menus, which are accessed when the Finder is the active window. When individual apps are open they have their own menu bars, although in a lot of cases these are similar to the standard Menu bar, particularly for the built-in macOS Big Sur apps such as Calendar, Contacts and Notes.

- **Apple menu**. This is denoted by a translucent gray apple and contains general information about the computer, a preferences option for changing the functionality and appearance of your MacBook, and options for closing down the computer.

- **Finder menu**. This contains preference options for amending the functionality and appearance of the Finder, and also options for emptying the Trash and accessing other apps (under the **Services** option).

- **File menu**. This contains common commands for working with open documents, such as opening and closing files, creating aliases, moving to the Trash, ejecting external devices, and burning discs.

- **Edit menu**. This contains common commands that apply to the majority of apps used on the MacBook. These include Undo, Cut, Copy, Paste, Select All and Show the contents of the Clipboard; i.e. items that have been cut or copied.

- **View menu**. This contains options for how windows and folders are displayed within the Finder and for customizing the Finder toolbar. This includes showing or hiding the Finder sidebar, and selecting view options for the size at which icons are displayed within Finder windows.

- **Go menu**. This can be used to navigate around your computer. This includes moving to your Recents folder, your Home folder, your Applications folder, and recently-accessed folders.

- **Window menu**. This contains commands to organize the currently-open apps and files on your Desktop.

- **Help menu**. This contains the MacBook Help files, which contain information about all aspects of macOS Big Sur.

Beware

An external Apple USB SuperDrive is required for burning CDs or DVDs on a MacBook, as they do not have an internal CD/DVD drive.

100

6 Navigating in macOS

macOS Big Sur uses Multi-Touch Gestures for navigating around your apps and documents. This chapter looks at how to use these to get around.

Navigating with MacBooks

macOS Big Sur on the MacBook uses a number of physical gestures to navigate around the operating system, folders and files. This involves a much greater reliance on swiping on a trackpad or adapted mouse; techniques that have been imported from the iPhone and the iPad. These are known as Multi-Touch Gestures and work most effectively with the trackpad on MacBooks. Other devices can also be used to perform Multi-Touch Gestures:

- **A Magic Trackpad**. This is an external trackpad that works wirelessly via Bluetooth, but there should not really be any need for one with a MacBook.

- **A Magic Mouse**. This is an external mouse that works wirelessly via Bluetooth.

These devices, and the trackpad, work using a swiping technique with fingers moving over their surface. This should be done with a light touch; it is a gentle swipe, rather than any pressure being applied to the device.

The trackpads and Magic Mouse do not have any buttons in the same way as traditional devices. Instead, specific areas are clickable so that you can still perform left- and right-click operations:

1 Click on the bottom-left corner for traditional left-click operations

2 **Ctrl + click** on the bottom-right corner for traditional right-click operations

Don't forget

The MacBook range all have Force Touch technology on the trackpad. This still allows for the usual range of Multi-Touch Gestures but can also be used for other functions, depending on the amount of pressure that is applied to the trackpad. Settings for this can be applied in the **Trackpad** section within **System Preferences**.

Pointing and Clicking

A trackpad or Magic Mouse can be used to perform a variety of pointing and clicking tasks.

1 Tap with one finger in the middle of the trackpad or Magic Mouse to perform a single-click operation; e.g. to click on a button or click on an open window

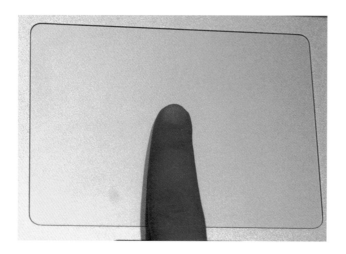

The Magic Mouse does not have separate buttons for different functions. Instead, the whole top of the Magic Mouse can be used to click on, or swipe on.

2 Tap with two fingers in the middle of the trackpad or Magic Mouse to access any contextual menus associated with an item (this is the equivalent of the traditional right-click with a mouse)

...cont'd

3 Highlight a word or phrase and double-tap with three fingers to see look-up information for the selected item. This is frequently a dictionary definition but it can also be a Wikipedia entry

Beware

If you have too many functions set using the same number of fingers, some of them may not work. See pages 114–116 for details about setting preferences for Multi-Touch Gestures.

4 Move over an item and drag with three fingers to move the item around the screen

macOS Scroll Bars

In macOS Big Sur, scroll bars in web pages and documents are more reactive to the navigational device being used on the computer. By default, with a trackpad or a Magic Mouse, scroll bars are only visible when scrolling is actually taking place. However, if a mouse is being used they will be visible permanently, although this can be changed for all devices. To perform scrolling with macOS Big Sur:

1 Scroll around a web page or document by swiping up or down on a trackpad or a Magic Mouse. As you move up or down, the scroll bar appears

Check **On** the **Always** option in Step 3 if you prefer to have the scroll bars visible at all times.

2 When you stop scrolling, the bar disappears to allow an optimum viewing area for your web page or document

3 To change the scroll bar options, select **System Preferences** > **General** and select the required setting under **Show scroll bars:**

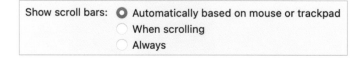

Don't forget

Don't worry if you cannot immediately get the hang of Multi-Touch Gestures. It takes a bit of practice to get the correct touch and pressure on the trackpad or Magic Mouse.

Scrolling and Zooming

One of the common operations on a computer is scrolling on a page, whether it is a web page or a document. Traditionally, this has been done with a mouse and a cursor. However, using a trackpad or Magic Mouse you can now do all of your scrolling with your fingers. There are a number of options for doing this:

Scrolling up and down

To move up and down web pages or documents, use two fingers on the trackpad and swipe up or down. The page moves in the opposite direction to the one in which you are swiping; i.e. if you swipe up, the page moves down, and vice versa:

1 Open a web page

2 Position two fingers in the middle of the trackpad

3 Swipe them up to move down the page

Don't forget

When scrolling up and down pages, the gesture moves the page the opposite way; i.e. swipe down to move up the page, and vice versa.

4 Swipe them down to move up a page

...cont'd

Zooming in and out

To zoom in or out on web pages or documents:

1 To zoom in, position your thumb and forefinger in the middle of the trackpad

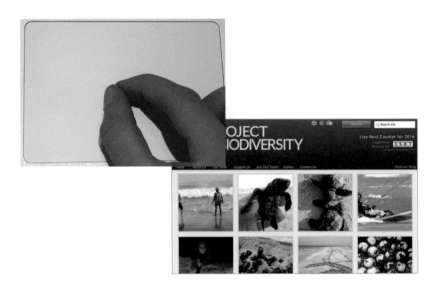

Pages can also be zoomed in on by double-tapping with two fingers. Double-tap again with two fingers to zoom back out.

2 Spread them outwards to zoom in on a web page or document

3 To zoom out, position your thumb and forefinger at opposite corners of the trackpad

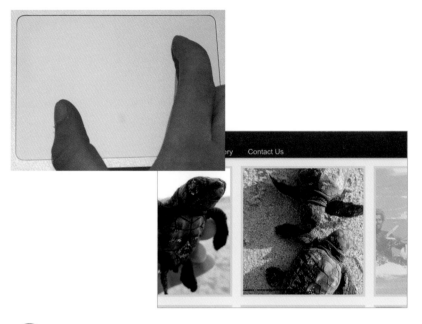

There is a limit to how far you can zoom in or out on a web page or document, to ensure that it does not distort the content too much.

4 Swipe them into the center of the trackpad to zoom out

...cont'd

Moving between pages
With Multi-Touch Gestures it is possible to swipe between pages within a document. To do this:

1 Position two fingers to the left or right of the trackpad

2 Swipe to the opposite side of the trackpad to move through the document

See pages 120-121 for details about using full-screen apps.

Moving between full-screen apps
In addition to moving between pages by swiping, it is also possible to move between different apps when they are in full-screen mode. To do this:

1 Position three fingers to the left or right of the trackpad

2 Swipe to the opposite side of the trackpad to move through the available full-screen apps

Showing the Desktop

To show the whole Desktop, regardless of how many files or apps are open:

1 Position your thumb and three fingers in the middle of the trackpad

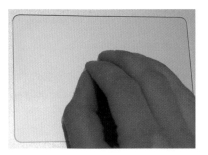

2 Swipe to the opposite corners of the trackpad to display the Desktop

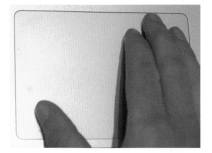

The Desktop can also be displayed, with any open items minimized around the side of the screen, by pressing the **Fn** + **F11** keys on the keyboard.

3 The Desktop is displayed, with all items minimized around the side of the screen

Mission Control and Spaces

Mission Control is a function in macOS Big Sur that helps you organize your open apps, full-screen apps and documents. It also enables you to quickly view the Desktop. Within Mission Control there are also Spaces, where you can group together similar types of documents. To use Mission Control:

1 Click on this button on the Dock, or

2 Swipe upwards with three fingers on the trackpad or Magic Mouse (or press **F3** on the keyboard)

 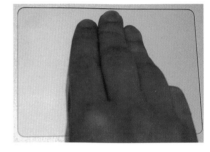

3 All open files and apps are visible via Mission Control

Don't forget

Click on a window in Mission Control to access it and exit the Mission Control window.

Beware

Any apps or files that have been minimized or closed do not appear within the main Mission Control window. Instead, they are located to the right of the dividing line on the Dock.

4 Move the cursor over the top of the Mission Control window to view the different Spaces and any apps in full-screen mode

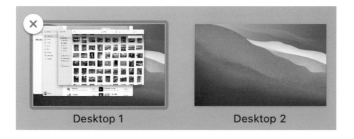

Desktop 1   Desktop 2

Preferences for Spaces can be set within the Mission Control system preferences.

Spaces

The top level of Mission Control contains Spaces, which are areas into which you can group certain apps; e.g. the Apple productivity apps such as Pages and Numbers. This means that you can access these apps independently from every other open item. This helps organize your apps and files. To use Spaces:

Create different Spaces for different types of content; e.g. one for productivity and one for entertainment.

1 Move the cursor over the top right-hand corner of Mission Control and click on the **+** symbol

2 A new **Space** is created along the top row of Mission Control

Desktop 2

When you create a new Space it can subsequently be deleted by moving the cursor over it and clicking on the cross at the left-hand corner. Any items that have been added to this Space are returned to the default Desktop Space.

3 Drag apps onto the Space. This can be accessed by clicking on the Space within Mission Control, and all of the apps that have been placed here will be available

Multi-Touch Preferences

Some Multi-Touch Gestures only have a single action, which cannot be changed. However, others have options for changing the action for a specific gesture. This is done within the trackpad preferences, where a full list of Multi-Touch Gestures is shown.

Point & Click preferences

1 Access System Preferences and click on the **Trackpad** button

2 Click on the **Point & Click** tab

3 The actions are described on the left, with a graphical explanation on the right

When setting Multi-Touch preferences try to avoid having too many gestures using the same number of fingers, in case some of them override the others.

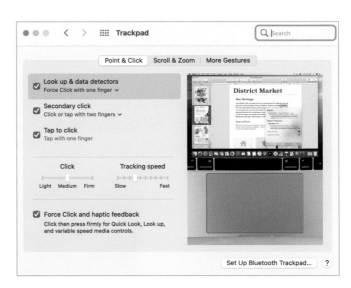

4 If there is a Down arrow next to an option, click on it to change the way an action is activated

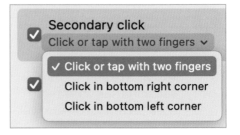

Scroll & Zoom preferences

1. Click on the **Scroll & Zoom** tab

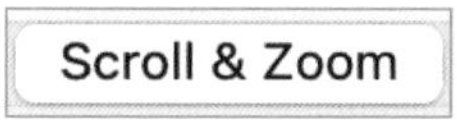

2. The actions are described on the left, with a graphical display on the right

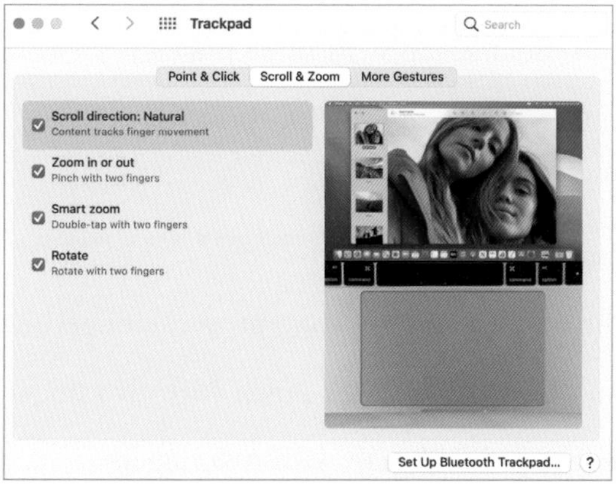

Hot tip

Experiment with the Multi-Touch Gestures in the trackpad system preferences, so that you become confident using them for when they are required.

More Gestures preferences

1. Click on the **More Gestures** tab

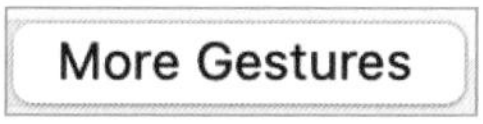

2. The actions are described on the left, with a graphical display on the right

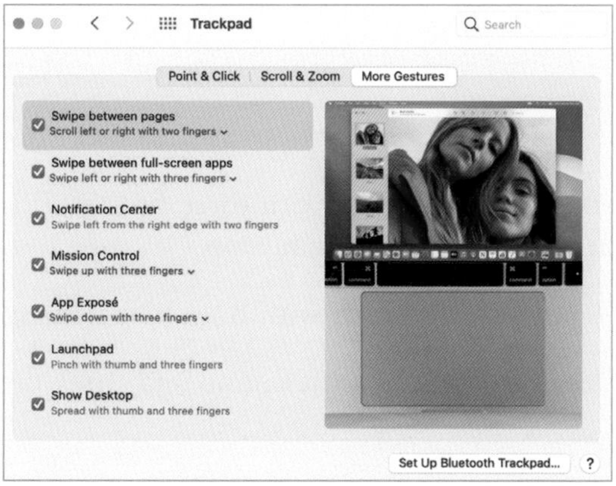

...cont'd

Multi-Touch Gestures
The full list of Multi-Touch Gestures, with their default actions, is:

Point & Click

- Tap to click – tap with one finger.
- Secondary click – click or tap with two fingers.
- Look up – double-tap with three fingers.
- Three-finger drag – move with three fingers.

Scroll & Zoom

- Scroll direction: natural – content tracks finger movement.
- Zoom in or out – spread or pinch with two fingers.
- Smart zoom – double-tap with two fingers.
- Rotate – rotate with two fingers.

More Gestures

- Swipe between pages – scroll left or right with two fingers.
- Swipe between full-screen apps – swipe left or right with three fingers.
- Access the Notification Center – swipe left with two fingers from the right-hand edge of the trackpad or Magic Trackpad.
- Access Mission Control – swipe up with three fingers.
- App Exposé – swipe down with three fingers. This displays the open windows for a specific app.
- Access Launchpad – pinch with thumb and three fingers.
- Show Desktop – spread with thumb and three fingers.

Hot tip

The Notification Center can also be accessed by clicking on the day and time icon in the top right-hand corner of the Desktop.

7 Working with Apps

Apps, or applications, are the programs with which you start putting macOS to use. This chapter looks at accessing the apps on your MacBook, and some of the functionality when using them. It also covers some of the most commonly-used apps, and looks at finding and downloading new ones from the online Apple App Store.

Launchpad

Even though the Dock can be used to store shortcuts to your applications, it is limited in terms of space. The full set of applications on your MacBook can be found in the Finder (in the Applications folder), but macOS Big Sur has a feature that allows you to quickly access and manage all of your applications. These include the ones that are pre-installed on your MacBook, and also any that you install yourself or download from the Apple App Store. This feature is called Launchpad. To use it:

Hot tip

If the apps take up more than one screen, swipe from right to left with two fingers to view the additional pages, or click on the dots at the bottom of the window.

Don't forget

To launch an app from within Launchpad, click on it once.

1 Click once on this button on the Dock

2 All of the apps (applications) are displayed

3 Similar types of apps can be grouped together in individual folders. By default, the utilities are grouped in this way

4 To create a group of similar apps, drag the icon for one over another

5 The apps are grouped together in a folder and Launchpad gives it a name, based on the type of apps within the folder

6 To change the name, click on it once and overtype it with the new name

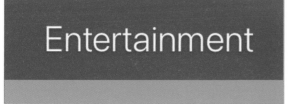

Most of the system apps (i.e. the ones that come pre-installed on your MacBook) cannot be removed from the Launchpad.

7 The folder appears within the **Launchpad** window

8 To remove an app, click and hold on it until it starts to jiggle and a cross appears. Click on the cross to remove it

Full-screen Apps

When working with apps we all like to be able to see as much of a window as possible. With macOS Big Sur this is possible with the full-screen option. This allows you to expand an app with this functionality so that it takes up the whole of your monitor or screen with a minimum of toolbars visible. Some apps have this functionality, but some do not. To use full-screen apps:

1 By default, an app appears on the Desktop with other windows behind it

If the button in Step 2 is not visible then the app does not have the full-screen functionality.

2 Click on this button at the top left-hand corner of the app's window

3 The app is expanded to take up the whole window. The main Apple Menu bar and the Dock are hidden

4 To view the main Menu bar, move the cursor over the top of the screen

5 You can move between all full-screen apps by swiping with three fingers left or right on the trackpad

For more information about navigating with Multi-Touch Gestures, see pages 102-116.

6 Move the cursor over the top left-hand corner of the screen and click on this button to close the full-screen functionality

7 In Mission Control, all of the open full-screen apps are shown in the top row

Desktop 1   Safari   Photos   Desktop 2

Some of the app icons in macOS Big Sur have been redesigned to give them a more uniform appearance, and also to make them consistent with the corresponding app icons on the iPhone and iPad.

Apple's productivity suite of apps (Keynote, Numbers and Pages) is available, for free, for qualifying MacBooks. The apps can also be downloaded, for free, from the App Store. GarageBand for creating music and iMovie for editing videos are also available in the same way.

macOS Apps

The built-in macOS Big Sur apps can be accessed from the Launchpad, as shown on page 118, or from the Applications folder within the Finder.

Some of the built-in macOS Big Sur apps include:

- **App Store**. This can be used to access the online App Store for viewing the range of apps there, and downloading new ones as required.

- **Books**. This can be used to access the online Book Store and download ebooks that you want to read. Downloaded books are stored in the app.

- **Calendar**. This can be used to record events and set reminders as required.

- **Contacts**. This is the macOS Big Sur address book, and contacts can be added with a range of details, including address, phone and mobile/cell number.

- **FaceTime**. This can be used for video calls or voice calls to family and friends using a MacBook or a mobile Apple device such as an iPhone or an iPad.

- **Home**. This can be used to control compatible smart home devices; e.g. smart lighting and heating.

- **Mail**. This is the macOS Big Sur default email app for sending and receiving emails. Photos and videos can be attached to emails with the app.

- **Maps**. This can be used to look up locations worldwide and also find directions between two different locations.

- **Messages**. This can be used to send text messages, including adding photos, videos and emojis. See pages 126-127 for more details.

- **Music**. This can be used to download music from the iTunes Store and then play tracks and albums. See page 129 for details.

● **News**. This can display collated news stories from a selection of different sources.

● **Notes**. This is Apple's note-taking app, which can be used to add checkboxes and hand-drawn notes. See pages 130-131 for details.

● **Podcasts**. This can be used to listen to audio programs, which can be subscribed to with the app.

● **Photo Booth**. This is an app for creating funny and quirky photo effects.

● **Photos**. This is the MacBook photo app and can be used to arrange, share and edit photos. See page 128 for details.

● **Preview**. This can be used to view files in a range of file formats such as JPEGs and GIFs for images, and also PDF files.

● **Reminders**. This can be used to set reminders on your MacBook. At the required time you will be alerted to the reminder.

● **Safari**. This is the macOS-specific web browser that supports tabbed browsing. See pages 124-125 for details about using Safari.

● **Siri**. This is Apple's digital voice assistant that is now available for macOS Big Sur. See pages 46-48 for details.

● **Stocks**. This can be used to display stock market information.

● **TV**. This can be used to access and download movies and TV shows from the Apple TV store. It can also be used to join the subscription channel Apple TV+.

● **Voice Memos**. This can be used to record voice memo, directly on your MacBook.

123

Safari App

Safari is a web browser that is designed specifically to be used with macOS. It is similar in most respects to other browsers, and works seamlessly with Big Sur.

Hot tip

If the Sidebar is not visible, select **View** from the Menu bar and click on the **Show Sidebar** option. From this menu you can also select or deselect items such as showing the Bookmarks Sidebar, the Tab Bar, the Status Bar and the Reading List Sidebar.

Hot tip

Safari uses a Smart Search box, so you can use the same box to search for a web address or use a general search keyword and click on one of the results.

Hot tip

Select **Safari** > **Preferences** from the Safari Menu bar to specify settings for the way Safari operates and displays web pages.

1 All of the controls are at the top of the browser

Sidebar button   Address/Search bar (Smart Search)   Tabs

2 Click on this button to open a new tab. This will open in the Top Sites window, from where a new website can be opened, or the Smart Search box can be used

3 Click on this button to view all of the currently-open tabs, as thumbnails

4 Click on this button to **Share** a web page, via email, messages or social media (or add a page as a bookmark, or to a Reading List)

5 Click on this button to view the Safari **Sidebar**, containing bookmarks, Reading List and shared links to subscribed feeds

Safari Start page

When you first open Safari, or open a new tab, the Start page is displayed. This can contain a selection of items that can be set within the Safari preferences. The options include favorites, frequently-visited pages and items selected by Siri. To use the Start page:

The Safari Start page has been updated in macOS Big Sur.

1 The Start page is displayed when Safari is opened, or when a new tab is opened (unless specified differently, in **Safari** > **Preferences** > **General** from the Safari Menu bar)

Hot tip

Ctrl + **click** on the Start page and click on the **Choose Background** option to select a background image for the Start page, from the Finder.

2 Swipe up the Start page to view the available options

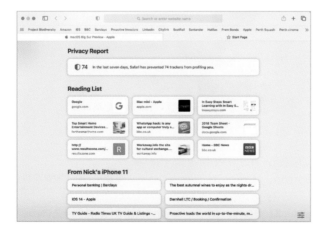

Messaging

The Messages app enables you to send text messages (iMessages) to other macOS users or those with an iPhone or iPod Touch using iOS, or an iPad using iPadOS. It can also be used to send photos and videos, and to make FaceTime calls. To use Messages:

The Messages app has been updated in macOS Big Sur.

1 Click on this icon on the Dock

2 Click on this button to start a new conversation

iMessages are specific to Apple and sent over Wi-Fi. SMS (Short Message Service) messages are usually sent between cellular devices, such as smartphones, that are contracted to a compatible mobile services provider. SMS messages can also be sent with the Messages app, to Apple or non-Apple users.

3 Click on this button and select a contact (these will be from your Contacts app).

To send an iMessage, the recipient must have an Apple ID

4 The person with whom you are having a conversation is displayed in the left-hand panel. The conversation continues down the right-hand panel

Audio messages can also be included in an iMessage. Click on this icon to the right of the text box, and record your message:

5 Click in the text box at the bottom of the main panel to write a message, and press **Return** on the keyboard to send

6 Click on this button at the right-hand side of the text box to access a range of emojis (small graphical symbols)

7 Click on an emoji to add it to a message

8 Scroll left and right on the window to view all of the emojis, or click on the bottom toolbar to move through the categories

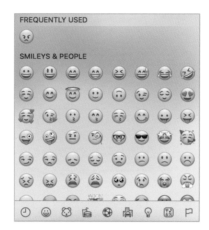

Messages in macOS also supports Tapback, whereby you can add an icon to a message as a quick reply. To do this, click and hold on a message text and click on an icon.

9 Some emojis of people have options to select different styles. Click and hold on a relevant item to view the options. Click on one to add it to a message

Pinning conversations

It is possible to keep your favorite conversations at the top of the left-hand panel by pinning them there. To do this:

1 **Ctrl + click** on a conversation in the left-hand panel and click on the **Pin** button

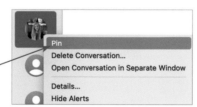

Pinning conversations is a new feature in macOS Big Sur.

2 The conversation is pinned at the top of the panel

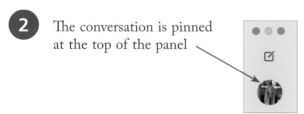

Photos App

The Photos app can be used to view photos according to Years, Collections, Moments, Memories or at full size. This enables you to view your photos according to dates and times at which they were taken. They can also be edited and shared in the Photos app:

1 Click here on the top toolbar to view **All Photos**, or by **Years**, **Months** and **Days**

Hot tip

Photos can be imported into the Photos app by selecting **File** > **Import** from the Menu bar and navigating to the required location within the Finder.

128

2 Double-click on a thumbnail to view it at full size

3 In full-size mode, click on the **Edit** button to access the editing options

Edit

4 Click on the **Memories** button in the left-hand sidebar to view collections of photos that are created by the app

Memories

Starting with the Music App

Music is a major part of the Apple ecosystem, and for many years iTunes has been integral to this. However, in macOS Big Sur, the iTunes app has been removed and replaced with a dedicated Music app. (Other forms of media content – e.g. movies and TV shows – also now have their own dedicated apps – e.g. the TV app or the Podcast app.) To use the Music app:

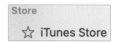

The iTunes Store is still available, from the Music app: click on the **iTunes Store** button in the left-hand sidebar.

1 Click on the **Music** app on the Dock

2 Under the **Library** heading in the left-hand sidebar, click on a category to view the available songs; e.g. by **Artists**

3 Click on a song or artist to view specific songs. Click on one to play it

4 Use these controls to, from left to right: shuffle songs; play previous; play/pause; play next; and repeat

By default, songs are stored in the iTunes Library and streamed over Wi-Fi to play them on your MacBook. However, they can also be downloaded to your MacBook by tapping on this button, so that they can be played at any time, even when you are offline:

5 Drag this slider to change the volume

129

Notes App

It is always useful to have a quick way of making notes of everyday things such as shopping lists, recipes or packing lists for traveling. With macOS Big Sur, the Notes app is perfect for this task. To use it:

Don't forget

Enable iCloud for Notes so that all of your notes will be backed up, and also available on all of your iCloud-enabled devices. You will also be able to access them by signing in to your account with your Apple ID at **www.icloud.com**

1 Click on this icon on the Dock, or in the Launchpad

2 The right-hand panel is where the note is created. The middle panel displays a list of all notes

3 Click on this button on the top toolbar to view the Notes Homepage as a gallery of thumbnails, rather than the default list option. This makes it easier to identify specific lists and their content

4 Click on this button to add a new note

Don't forget

The first line of a note becomes its heading in the notes panel.

5 As more notes are added, the most recent appears at the top of the list in the middle panel (below any notes that have been pinned). Click on the new note to add text and edit formatting options (see next page)

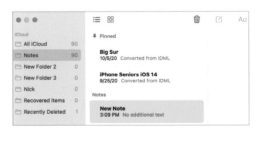

Formatting notes

In macOS Big Sur there are a number of formatting options for the Notes app:

1 Enter a line of text, and click on this button to add a check button

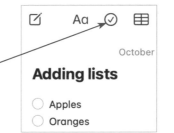

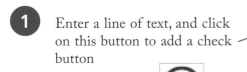

2 Click on the check button to add a check mark, to indicate that an item has been completed

3 Highlight a piece of text, and click on this button to access formatting options for it

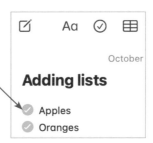

4 Click on this button on the Notes toolbar to add photos and videos, sketches, scans, maps, websites, sound clips and documents to a note:

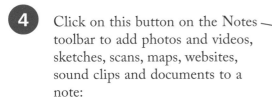

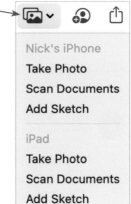

Click on this button to add a table to a note:

Click on the **Share** button to share a note with other people, via a range of other apps.

Click on this button to invite other people to have access to the note so they can view it and edit it:

Content can be added to notes from a range of other apps, such as Safari or the Photos app, by clicking on the **Share** button in the relevant app and selecting **Notes** as the option.

macOS on a MacBook can be used for phone call forwarding from your iPhone. Both devices need to have Wi-Fi turned on and be signed in to the same iCloud account. On your MacBook, select **FaceTime** > **Preferences** from the FaceTime menu and check On **Calls from iPhone** (from the Settings tab). On your iPhone, access **Settings** > **Phone** > **Calls on Other Devices** and drag the **Allow Calls on Other Devices** On, and select a device under the **Allow Calls On** section.

When a call is connected, click on this button and click on the **Add Person** button to add someone else to the call and create a group call:

FaceTime

FaceTime is an app that can be used to make video and audio calls to other Macs, iPhones, iPads and iPod Touches. To use FaceTime on your MacBook you must have an in-built FaceTime camera, or use a compatible external one. To use FaceTime:

1 Click on this icon on the Dock

2 You need an Apple ID to use FaceTime. Click here to enter the name, email address or cell number of the person you want to contact. Click on the video icon to make the call

3 When a call is connected, the other person's video feed is in the main window and your own is shown as a thumbnail

4 Move the cursor over the FaceTime screen to access the control buttons for, from left to right: viewing the details panel; muting a call; ending a call; turning off your video feed; and making the window full-screen

Accessing the App Store

The Apple App Store is an online facility where you can download and buy new apps. These cover a range of categories such as productivity, business and entertainment. When you select or buy an app from the App Store, it is downloaded automatically by Launchpad and appears here next to the rest of the apps.

To buy apps from the App Store you need to have an Apple ID. If you have not already set this up, it can be done when you first access the App Store. To use the App Store:

1 Click on this icon on the Dock or within the Launchpad

2 The Homepage of the App Store contains the current top featured apps

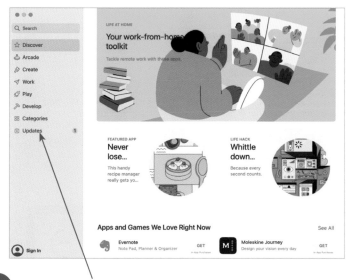

3 Your account information and quick links to other areas of the App Store are listed in the left-hand sidebar

The App Store is an online function so you will need an internet connection to access it.

You can set up an Apple ID when you first set up your MacBook, or you can do it when you register for the App Store or another online service, such as Messages.

Downloading Apps

The App Store contains a wide range of apps: from small, fun apps to powerful productivity ones. However, downloading them from the App Store is the same regardless of the type of app. The only differences are whether they need to be paid for or not and the length of time they take to download. To download an app from the App Store:

Hot tip

When downloading apps, start with a free one first so that you can get used to the process before you download paid-for apps.

1 Browse through the App Store until you find the required app

HabitMinder
Good habit tracker & motivator   GET

2 Click on the app to view a detailed description about it

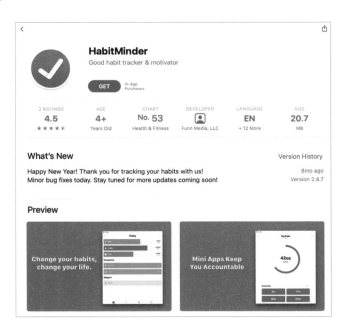

3 Click on the button underneath the app icon to download it. If there is no charge for the app the button will say **Get**

4 If there is a charge for the app, the button will say **Buy App**

5 Enter your Apple ID account details and click on the **Buy** button to download the app

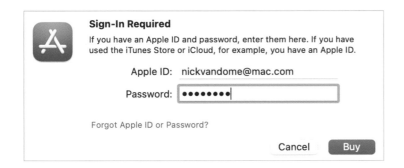

Sign-In Required

If you have an Apple ID and password, enter them here. If you have used the iTunes Store or iCloud, for example, you have an Apple ID.

Apple ID: nickvandome@mac.com

Password: ••••••••

Forgot Apple ID or Password?

Cancel   Buy

Depending on their size, different apps take differing amounts of time to be downloaded.

135

6 The progress of the download is displayed in a progress bar underneath the Launchpad icon on the Dock

7 Once it has been downloaded, the app is available within Launchpad

Siri   Mission Control

Notes   HabitMinder

As you download more apps, additional pages will be created within the Launchpad to accommodate them.

Finding Apps

There are thousands of apps in the App Store and sometimes the hardest task is locating the ones you want. However, there are a number of ways in which finding apps is made as easy as possible:

1 Click on the **Discover** button

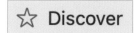

2 The main window has a range of suggested apps. Scroll down the page to see a panel with the current **Top Free** apps and games

3 Underneath this is a list of the current **Top Paid** apps and games

4 Click on the **See All** button to view the full list of Top Charts apps

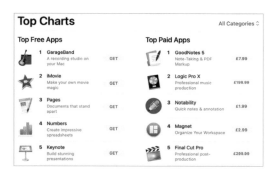

Don't forget

The Top Charts has sections for paid-for apps and free ones.

5 Click on the buttons in the left-hand panel to view the available apps related to these headings

6 Each heading contains the relevant apps

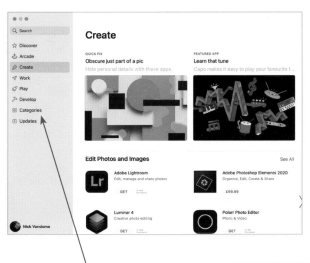

Another way to find apps is to type a keyword into the Search box at the top left-hand corner of the App Store window.

7 Click on the **Categories** button

88 Categories

8 Browse through the apps by specific categories, such as Business, Entertainment and Finance

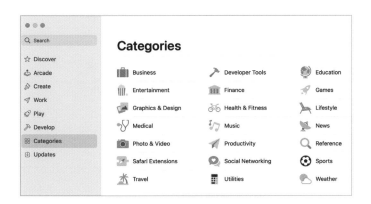

Managing Your Apps

Once you have bought apps from the App Store you can view details of ones you have purchased and also install updated versions of them. To view your purchased apps:

1 Click on your own account name, in the bottom left-hand corner

2 Details of your purchased apps are displayed (including those that are free)

Updating apps

Improvements and fixes are being developed constantly, and these can be downloaded to ensure that all of your apps are up-to-date:

1 When updates are available this is indicated by a red circle on the App Store icon on the Dock

2 Click on the **Updates** button

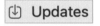

3 Information about the update is displayed next to the app that is due to be updated

4 Click on the **Update** button to update an individual app

5 Click on the **Update All** button to update all of the apps that are due to be updated

8 Sharing macOS

This chapter looks at how to set up different user accounts and use Screen Time.

Adding Users

macOS enables multiple users to access individual accounts on the same computer. If there are multiple users (i.e. two or more for a single machine), each person can sign on individually and access their own files and folders. This means that each person can log in to their own settings and preferences. All user accounts can be password-protected, to ensure that each user's environment is secure. To set up multiple user accounts:

Don't forget

Every computer with multiple users has at least one main user, also known as an administrator. This means that they have greater control over the number of items they can edit and alter. If there is only one user on a computer, they automatically take on the role of administrator. Administrators have a particularly important role to play when computers are networked together. Each computer can potentially have several administrators.

Don't forget

Each user can select their own icon or photo of themselves.

1 Click on the **System Preferences** icon on the Dock

2 Click on the **Users & Groups** icon

3 The information about the current account is displayed. This is your own account, and the information is based on details you provided when you first set up your MacBook

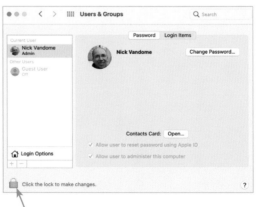

4 Click on this icon to enable new accounts to be added (the padlock needs to be open)

5 Click on the **+** button to add a new account

6 Enter the details for the new account holder

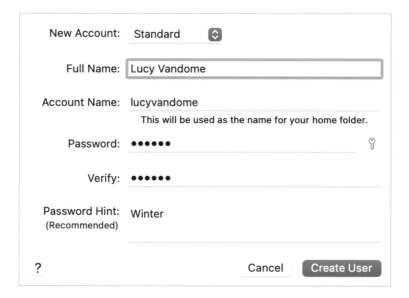

New Account: Standard

Full Name: Lucy Vandome

Account Name: lucyvandome
This will be used as the name for your home folder.

Password: ••••••

Verify: ••••••

Password Hint: Winter
(Recommended)

? Cancel Create User

7 Click on the **Create User** button

Create User

8 The new account is added to the list in the Users & Groups window, under **Other Users**

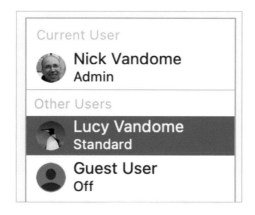

Current User

Nick Vandome
Admin

Other Users

Lucy Vandome
Standard

Guest User
Off

At Step 6, choose the type of account from the drop-down list. An **Administrator** account is one that allows the user to make system changes, and add or delete other users; a **Standard** account allows the user to use the functionality of the MacBook, but not change system settings; and **Sharing Only** is an account that allows guests to log in temporarily, without a password – when they log out, all files and information will be deleted from the guest account.

By default, you are the administrator of your own MacBook. This means that you can administer other user accounts.

Deleting Users

Once a user has been added, their name appears on the list in the Users & Groups dialog box. It is then possible to edit the details of a particular user or delete them altogether. To do this:

Beware

Always tell other users if you are planning to delete them from the system. Don't just remove them and then let them find out the next time they try to log in. If you delete a user, their personal files can be left untouched and can still be accessed (by selecting **Don't change the home folder** in Step 3).

1 Within **Users & Groups**, unlock the settings as shown on page 140, then select a user from the list

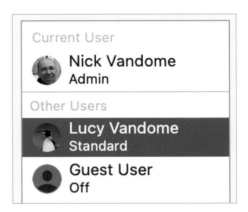

2 Click here to remove the selected person's user account

3 A warning box appears, to check if you really do want to delete the selected user. If you do, select the required option and click on the **Delete User** button

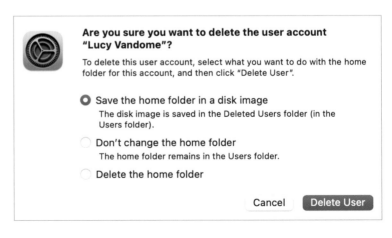

Fast User Switching

If there are multiple users on your macOS system, it is useful to be able to switch between them as quickly as possible. When this is done, the first user's session is retained so that they can return to it if required. To switch between users:

Unlock the settings before you start (see Step 4 on page 140).

1 In the Users & Groups window, click on the **Login Options** button

2 Check **On** the **Show fast user switching menu as** box, then close the window

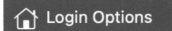

3 At the top right of the screen, click on the current user's name

Nick Vandome

When you switch between users, the first user remains logged in and their current session is retained intact.

4 Click on the name of another user

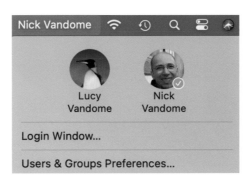

5 Enter the relevant password (if required)

Users can sign in from the Lock screen with a password by clicking on their own icon/name on the screen and entering the relevant details. If fast user switching is not used, each user has to log out before the next one can sign in.

6 Click on this button to log in

Screen Time

The amount of time that we spend on our digital devices is a growing issue in society, and steps are being taken to let us see exactly how much time we are spending looking at our computer screens. In macOS Big Sur, a range of screen-use options can be monitored with the Screen Time feature. To use this:

1 Select **System Preferences** > **Screen Time**

Screen Time

2 Click on the **Options** button

··· Options

3 Click on the **Turn On...** button to activate Screen Time

Screen Time for this Mac: **Off**   Turn On...

☐ Share across devices
You can enable this on any iPhone, iPad, or Mac signed in to iCloud to report your combined screen time.

☐ Use Screen Time Passcode   Change Passcode...
Use a passcode to secure Screen Time settings, and to allow for more time when limits expire.

Don't forget

The amount of screen usage time is updated automatically in the window in Step 4.

4 Click on the **App Usage** button in the left-hand sidebar to view the overall screen usage. Click here to select a different time period to view

⎍ App Usage

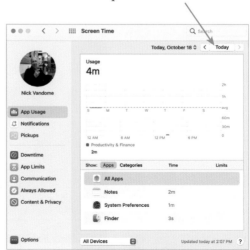

Downtime

This can be used to specify times when the computer and its apps cannot be accessed:

All of the Screen Time options can be accessed from the left-hand sidebar of the main window.

1 Check **On** the **Every Day** button to apply time limits for a whole week

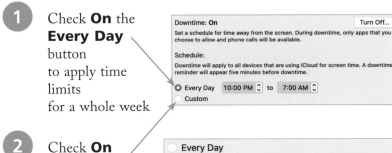

2 Check **On** the **Custom** button to apply time limits for individual days

App Limits

This can be used to limit the amount of time that certain categories of apps can be accessed:

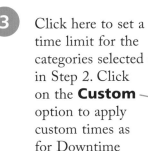

In general, the categories for app limits relate to the categories under which the apps are listed in the App Store.

1 Click on the **+** button at the bottom of the **App Limits** window

2 Click on any category to which you want limits to apply

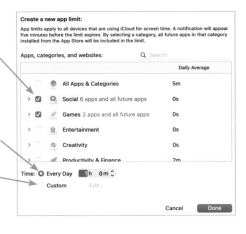

3 Click here to set a time limit for the categories selected in Step 2. Click on the **Custom** option to apply custom times as for Downtime

...cont'd

Communication

This can be used to specify who can contact you during Screen Time:

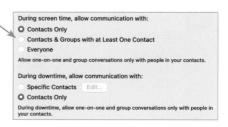

1 Click here to select who can contact you during Screen Time and also Downtime, from contacts or everyone

The Communication Screen Time option is a new feature in macOS Big Sur.

Always Allowed

This can be used to allow the use of specified apps, regardless of other settings:

1 Check **On** the items that you want to allow, regardless of any other Screen Time settings

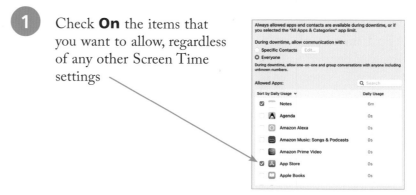

Don't forget

Click on the **Stores** tab under **Content & Privacy** to access options for content that is downloaded from the App Store, the Music app, the TV app and the Podcasts app. Selections can be made for age-appropriate content and also blocking content with explicit language. Click on the **Apps** tab in the Content & Privacy Restrictions window to allow the use of specific apps regardless of other privacy restrictions.

Content & Privacy

This can be used to add restrictions based on the type of content being viewed:

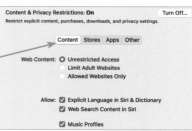

1 Click on the **Content** tab to restrict content based on criteria such as adult content and explicit language

9 MacBook Networking

This chapter looks at how to use your MacBook to create and work with networks, for sharing information.

Networking Overview

Before you start sharing files directly between computers, you have to connect them together. This is known as networking and can be done with two computers in the same room, or with thousands of computers in a major corporation. If you are setting up your own small network it will be known in the computing world as a Local Area Network (LAN). When setting up a network, there are various pieces of hardware that are initially required to join all of the required items together. Once this has been done, software settings can be applied for the networked items. Some of the items of hardware that may be required include:

- **A network card**. This is known as a Network Interface Card (NIC), and all recent Macs have them built in.

- **A wireless router**. This is for a wireless network, which is increasingly the most common way to create a network, via Wi-Fi. The router is connected to a telephone line, and the computer then communicates with it wirelessly.

- **An Ethernet port and Ethernet cable**. This enables you to make the physical connection between devices. Ethernet cables come in a variety of forms, but the one you should be looking for is the Cat7 type, as this allows for the fastest transfer of data. If you are creating a wireless network then you will not require these. The latest range of MacBooks do not have a dedicated Ethernet port, but an Ethernet to Thunderbolt adapter can be used.

- **A hub**. This is a piece of hardware with multiple Ethernet ports that enables you to connect all of your devices together and lets them communicate with each other. However, conflicts can occur with hubs if two devices try to send data through one at the same time.

- **A switch**. This is similar in operation to a hub but it is more sophisticated in its method of data transfer, thus allowing all of the machines on the network to communicate simultaneously, unlike a hub.

Once you have worked out all of the devices that you want to include on your network, you can arrange them accordingly. Try to keep the switches and hub within relative proximity of a power supply and, if you are using cables, make sure they are laid out safely.

Connecting to the internet is also another form of network connection.

Ethernet network

The cheapest and easiest way to network computers is to create an Ethernet network. This involves buying an Ethernet hub or switch, which enables you to connect several devices to a central point; i.e. the hub or switch. All Apple computers and most modern printers have an Ethernet connection, so it is possible to connect various devices, not just computers. Once all of the devices have been connected by Ethernet cables, you can then start applying network settings.

AirPort network

Another option for creating a network is using Apple's own wireless system, AirPort. This creates a wireless network, and there are two main options used by Apple computers: AirPort Express, using the IEEE 802.11n standard, which is more commonly known as Wi-Fi; and the newer AirPort Extreme, using the next-generation IEEE 802.11ac standard, which is up to five times faster than the 802.11n standard. Thankfully, AirPort Express and Extreme are also compatible with devices based on the older IEEE standards, 802.11b/g/n, so one machine loaded with AirPort Extreme can still communicate wirelessly with the older AirPort version.

One of the main issues with a wireless network is security, since it is possible for someone with a wireless-enabled machine to access your wireless network if they are within range. However, in the majority of cases the chances of this happening are fairly slim, although it is an issue about which you should be aware.

The basic components of a wireless network between Macs is an AirPort card (either AirPort Express or AirPort Extreme) installed in all of the required machines, and an AirPort base station that can be located anywhere within 150 meters of the AirPort-enabled computers. Once the hardware is in place, wireless-enabled devices can be configured by using the AirPort Setup Assistant utility found in the Utilities folder. After AirPort has been set up, the wireless network can be connected. All of the wireless-enabled devices should then be able to communicate with each other, without the use of a multitude of cables.

Wireless network

A wireless network can also be created with a standard wireless router, rather than using the AirPort option.

Don't forget

Another method for connecting items wirelessly is called Bluetooth. This covers much shorter distances than AirPort, and is generally used for items like printers and cell phones. Bluetooth devices can be connected by using the Bluetooth option in System Preferences.

Network Settings

Once you have connected the hardware required for a network, you can start applying the network settings that are required for connecting to the internet, for online access.

Beware

If you turn off the Wi-Fi function, this will disconnect you from your network and also the internet.

1 In **System Preferences**, click on the **Network** button

Network

2 For a wireless connection, click on the **Turn Wi-Fi On** button

Turn Wi-Fi On

3 Details of wireless settings are displayed

4 If you are already connected to a network, this will be shown in the **Status** section

Status: **Connected**   Turn Wi-Fi Off

Wi-Fi is connected to PLUSNET-TXJ5 and has the IP address 192.168.1.75.

Don't forget

If you are asked to join new networks, all of the available networks in range will be displayed, not just the one you may want to join.

5 The network name is shown here

Network Name: PLUSNET-TXJ5

6 Check **On** this box if you want to be notified before joining a new network

☑ **Ask to join new networks**

Known networks will be joined automatically. If no known networks are available, you will be asked before joining a new network.

Connecting to Bluetooth

The network settings can also be used to set up Bluetooth devices, such as wireless keyboards or speakers. To do this:

1 In the **Network** window on the previous page, click on the **Bluetooth PAN** button in the left-hand sidebar

2 Click on the **Set Up Bluetooth Device...** button in the main window

3 Click on the **Turn Bluetooth On** button

4 Turn on the Bluetooth device and ensure that it is in "pairing" mode; i.e. it can communicate with other Bluetooth devices and connect with them. Devices that are paired are listed under the **Devices** heading and can then be used with the MacBook; e.g. if music is being played on the MacBook, a Bluetooth speaker can be used to have the music directed there instead

Beware

Bluetooth devices have to be within a range of about 20 meters in order to be paired and communicate with another device. However, the closer the better.

Don't forget

Once a Bluetooth device has been paired with your MacBook, click on the **Connect** button so that it can be used with the MacBook.

151

Connecting to a Network

Connecting as a registered user

To connect as a registered user (usually as yourself when you want to access items on another one of your own computers):

Don't forget

The username and password used to connect to a networked computer as a registered user are the ones used to log in to your MacBook. There is also an option to connect using your Apple ID.

Don't forget

You can disconnect from a networked computer by ejecting it in the Finder in the same way as you would for a removable drive, such as a flashdrive (see page 32) by clicking on the **Eject** icon.

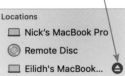

Locations
* Nick's MacBook Pro
* Remote Disc
* Eilidh's MacBook... ⏏

---

1 Click on the **Network** button on the Finder sidebar

🌐 Network

2 Click on the networked computer to which you want to connect

Locations
* 🖥 Eilidh's MacBook Pro

3 Click on the **Connect As...** button

Connect As...

4 Check **On** the **Registered User** button, and enter your username and password

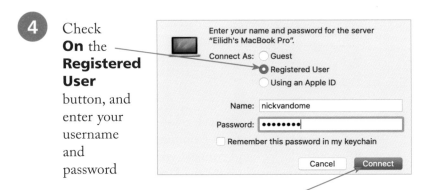
Enter your name and password for the server "Eilidh's MacBook Pro".

Connect As: ○ Guest
● Registered User
○ Using an Apple ID

Name: nickvandome
Password: ••••••••
☐ Remember this password in my keychain

Cancel   Connect

5 Click on the **Connect** button

6 The public folders and home folder of the networked computer are available to the registered user. Double-click on an item to view its contents

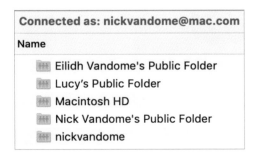
Connected as: nickvandome@mac.com

Name
* 👥 Eilidh Vandome's Public Folder
* 👥 Lucy's Public Folder
* 👥 Macintosh HD
* 👥 Nick Vandome's Public Folder
* 👥 nickvandome

Guest users

Guest users on a network are users other than yourself or other registered users, to whom you want to limit access to your files and folders. Guests only have access to a folder called the Drop Box in your own Public folder. To share files with guest users you have to first copy them into the Drop Box. To do this:

1 Create a file and select **File > Save** from the Menu bar

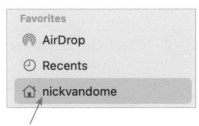

If another user is having problems accessing the files in your Drop Box, check the Permissions settings that have been assigned to the files. See page 186 for further details.

2 Navigate to your own home folder (this is created automatically by macOS and displayed in the Finder sidebar with your MacBook username)

3 Double-click on the **Public** folder

Public

The contents of the Drop Box can be accessed by other users on the same computer as well as by users on the network.

4 Double-click on the **Drop Box** folder

Drop Box

5 Save the file into the Drop Box

The Drop Box folder is not the same as the Dropbox app, which is a third-party app used for online storage and backing up.

...cont'd

Accessing a Drop Box
To access files in a Drop Box:

1 Access a networked computer as shown on page 152

2 Click on the **Connect As...** button in the Finder window

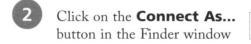

3 Check **On** the **Guest** button

4 Click on the **Connect** button

5 Double-click on a user's **Public Folder**

Nick Vandome's Public Folder

6 Double-click on the **Drop Box** folder to access the files within it

Drop Box

Beware

It is better to copy files into the Drop Box rather than moving them completely from their primary location.

Hot tip

Set permissions for how the Drop Box operates by selecting it in the Finder and **Ctrl + clicking** on it. Select **Get Info** from the menu, and apply the required settings under the **Sharing & Permissions** heading.

File Sharing

One of the main reasons for creating a network of two or more computers is to share files between them. On networked MacBooks, this involves setting them up so that they can share files and then access them.

Setting up file sharing

To set up file sharing on a networked MacBook:

1 Click on the **System Preferences** icon on the Dock

2 Click on the **Sharing** icon and click on the padlock to unlock the settings, as shown on page 140

Sharing

3 Check **On** the boxes next to the items you want to share (the most common items to share are files and printers)

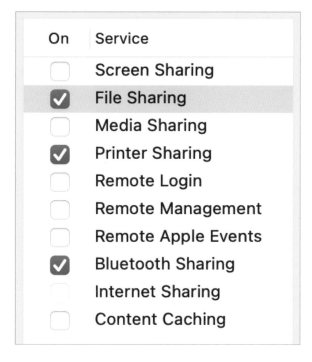

On	Service
☐	Screen Sharing
☑	File Sharing
☐	Media Sharing
☑	Printer Sharing
☐	Remote Login
☐	Remote Management
☐	Remote Apple Events
☑	Bluetooth Sharing
☐	Internet Sharing
☐	Content Caching

If no file-sharing options are enabled in the Sharing preferences window, no other users will be able to access your computer or your files, even on a network.

Networks can also be created, and items shared, between Macs and Windows-based PCs.

Sharing with AirDrop

Files can also be shared between Mac computers using the AirDrop feature. This enables the devices to connect wirelessly, and you can then share content between them. To do this:

1 In the Finder sidebar, click on the **AirDrop** button

2 Other users, or devices, in range with AirDrop are shown in the main Finder window

3 Drag content from another Finder window (or the Desktop) over the user icon to share it with them

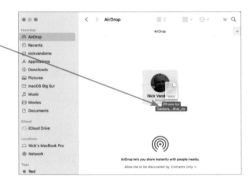

4 Content that has been shared from another device is initially placed in the **Downloads** folder in the Finder. It can then be copied from here and placed in another location, if required

10 MacBook Mobility

MacBooks are ideal for mobile working, for business or pleasure. This chapter looks at being mobile with your MacBook, including protecting it and security concerns.

Transporting Your MacBook

When you are going traveling, either for business or pleasure, your MacBook can be a very valuable companion. It can be used to download photographs from a digital camera, download movies from a digital video camera, keep a diary or business notes, and keep a record of your itinerary and important documents. Also, in many parts of the world it can access the internet via wireless hotspots so that you can view the web and send emails. However, when you are traveling with your MacBook it is sensible to transport this valuable asset in as safe and secure a way as possible. Some of the options include:

MacBook cases and sleeves

There is a range of MacBook cases and sleeves designed specifically for providing protection for the MacBook. They can be bought from the Apple website or Apple stores.

Metal case

If you are concerned that your MacBook may be in danger of physical damage when you are on the road, you may want to consider a more robust metal case. These are similar to those used by photographers and, depending on its size and design, you may also be able to include any photographic equipment.

Backpacks

A serious option for transporting your MacBook while you are traveling is a small backpack. This can either be a standard backpack or a backpack specifically designed for a MacBook. The latter is clearly a better option as the MacBook will fit more securely and there are also pockets designed for accessories.

Don't forget

A backpack for carrying a MacBook can be more comfortable than a shoulder bag, as it distributes the weight more evenly.

Keeping Your MacBook Safe

By most measures, MacBooks are valuable items. However, in a lot of countries around the world their relative value can be a lot more than it is to their owners: in some countries the value of a MacBook could easily equate to a month's, or even a year's, wages. Even in countries where their relative value is not so high they can still be seen as a lucrative opportunity for thieves. Therefore, it is important to try to keep your MacBook as safe as possible when you are traveling with it, either abroad or at home. Some points to consider in relation to this are:

- If possible, keep your MacBook with you at all times; i.e. transport it in a piece of luggage that you can carry rather than having to put it into a large case.

- Never hand over your MacBook, or any of your belongings, to any local who promises to look after them.

- If you do have to detach yourself from your MacBook, put it somewhere secure such as a hotel safe.

- When you are traveling, keep your MacBook as unobtrusive as possible. This is where a backpack carrying case can prove useful, as it is not immediately apparent that you are carrying a MacBook.

- Do not use your MacBook in areas where you think it may attract undue interest from the locals, particularly in obviously poor areas. For instance, if you are in a local café the appearance of a MacBook may create unwanted attention for you. If in doubt, wait until you get back to your hotel.

- If you are accosted by criminals who demand your MacBook, hand it over. No piece of equipment is worth suffering physical injury for.

- If you are abroad make sure your MacBook is covered by your travel insurance. If not, get separate insurance for it.

- Trust your instincts with your MacBook. If something doesn't feel right, don't do it.

Hot tip

Save your important documents onto a flashdrive, or an external hard drive, on a daily basis when you are traveling and keep this away from your MacBook. Alternatively, back up to iCloud when Wi-Fi is available. This way, you will still have these items if your MacBook is lost or stolen.

Temperature Extremes

Traveling consists of seeing a lot of different places and cultures, but it also invariably involves different extremes of temperature: a visit to the pyramids of Egypt can see the mercury in the upper reaches of the thermometer, while a trip to Alaska would encounter much colder conditions. Whether it is hot or cold, looking after your MacBook is an important consideration in extremes of temperature.

Heat

When traveling in hot countries, the best way of avoiding any heat damage to your MacBook is to prevent it from getting too hot in the first place:

- Do not place your MacBook in direct sunlight.

- Keep your MacBook insulated from the heat.

- Do not leave your MacBook in an enclosed space, such as a car. Not only can this get very hot, but the sun's power can be increased by the vehicle's glass.

Cold

Again, it is best to avoid your MacBook getting too cold in the first place, and this can be done by following similar precautions to those for heat. However, if your MacBook does suffer from extremes of cold, allow it to warm up to normal room temperature again before you try to use it. This may take a couple of hours but it will be worth the wait, rather than risking damaging the delicate computing elements inside.

If a MacBook gets too hot it could buckle the casing, making it difficult to close.

Wrap your MacBook in something white, such as a T-shirt or a towel, to insulate it against extreme heat.

Dealing with Water

Water is one of the greatest enemies of any electrical device, and MacBooks are no different. This is of particular relevance to anyone who is traveling near water with their MacBook, such as on a boat or ship, or using their MacBook near a swimming pool or a beach. If you are near water with your MacBook then you must bear the following in mind:

- **Avoid water**. The best way to keep your MacBook dry is to keep it away from water whenever possible. For instance, if you want to update your notes or download some photographs, then it would be best to do this in an indoor environment, rather than sitting near water.

- **Keep dry**. If you think you will be transporting your MacBook near water then it is a good precaution to protect it with some form of waterproof bag. There is a range of "dry-bags" that are excellent for this type of occasion and they remain waterproof even if fully immersed in water. These can be bought from outdoor suppliers.

- **Dry out**. If the worst does occur and your MacBook does get a good soaking, all is not necessarily lost. However, you will have to ensure that it is fully dried out before you try to use it again. Never turn it on if it is still wet.

Power Sockets

Different countries and regions around the world use different types of power sockets, and this is an issue when you are traveling with your MacBook. Wherever you are going in the world, it is vital to have an adapter that will fit the sockets in the countries you intend to visit. Otherwise, you will not be able to charge your MacBook battery.

There are over a dozen different types of plugs and sockets used around the world, with the four most popular being:

North America/Japan
This is a two-point plug and socket.
The pins on the plug are flat and parallel.

Continental Europe
This is a two-point plug and socket.
The pins are rounded.

Australasia/China/Argentina
This is a three-point socket that can accommodate either a two- or a three-pin plug. In a two-pin plug, the pins are angled in a V shape.

UK
This is a three-point plug and socket.
The pins are rectangular.

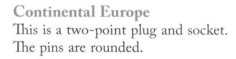

Hot tip

Power adapters can be bought for all regions around the world. There are also kits that provide all of the adapters together. These provide connections for anywhere worldwide.

Airport Security

Because of the increased global security following terrorist attacks, including those on planes or at airports, levels of airport security have been greatly increased around the world. This has implications for all travelers, and if you are traveling with a MacBook, this will add to the security scrutiny you will face. When dealing with airport security when traveling with a MacBook, there are some issues you should always keep in mind:

- Keep your MacBook with you at all times. Unguarded baggage at airports immediately raises suspicion, and it can make life very easy for thieves.

- Carry your MacBook in a small bag so you can take it on board as hand luggage. On no account should it be put in with your luggage that goes in the hold.

- X-ray machines at airports will not harm your MacBook. However, if anyone tries to scan it with a metal detector, ask them if they can inspect it by hand instead.

- Keep a careful eye on your MacBook when it goes through the X-ray conveyor belt, and try to be there at the other side as soon as it emerges. There have been some stories of people causing a commotion at the security gate just after someone has placed their MacBook on the conveyor belt. While everyone's attention (including yours) is distracted, an accomplice takes the MacBook from the conveyor belt. If you are worried about this you can ask for the security guard to hand-check your MacBook rather than putting it on the conveyor belt.

- Make sure the battery of your MacBook is fully charged. This is because you may be asked to turn on your MacBook to verify that it is just that, and not some other device disguised as a MacBook. This check has become increasingly common in recent years due to security threats, and for some countries, such as the US, digital devices have to be turned on.

- When you are on the plane, keep the MacBook in the storage area under your seat, rather than in the overhead locker, so you know where it is at all times. Also, it could cause a serious injury if it fell out of an overhead locker.

Beware

If there is any kind of distraction when you are going through airport security it could be because someone is trying to divert your attention in order to steal your MacBook.

Hot tip

When traveling through airport security, leave your MacBook in Sleep mode, so it can be powered up quickly if anyone needs to check that it works properly.

Some Apps for Traveling

When you are traveling with your MacBook you can use it for productivity tasks with the Apple suite of apps: Pages, Numbers and Keynote. In addition, Safari can be used to connect wirelessly to the web, and you can use the Music app for listening to music and the Photos app for your photos. There are also some other built-in apps that can be useful when you are traveling:

- **Notes**. Use this to create notes relating to your trip, ranging from Things to Pack lists to health information and travel details such as your itinerary. It is also a good option for keeping track of important items such as passport numbers.

- **Books**. Books are an ideal traveling companion, and with this app you do not have to worry about being weighed down by a lot of heavy volumes.

- **Maps**. This is the perfect app for researching cities abroad so that you can start to feel at home as soon as you arrive. You can view maps in standard or 3D satellite view, and also get directions between two locations.

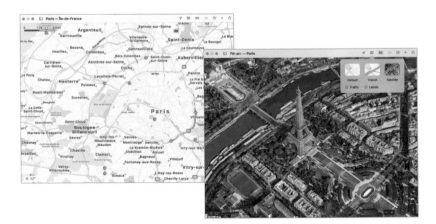

Hot tip

For some locations, Maps has an automated 3D tour of a city. If this is available for the location being viewed, a **Flyover Tour** button will appear. Click on this to begin the flyover tour.

- **FaceTime**. You can use your MacBook to send emails and text messages when you are away from home, but FaceTime allows you to see people too, with voice and video calls.

11 Battery Issues

Battery power is crucial to a MacBook. This chapter shows how to get the best from your battery.

Power Consumption

Battery life for each charge of MacBook batteries is one area that engineers have worked very hard on since MacBooks were first introduced. For the latest models of MacBooks, the average battery life for each charge is approximately 11-17 hours for wireless web use. However, this is dependent on the power consumption of the MacBook; i.e. how much power is being used to perform particular tasks. Power-intensive tasks will reduce the battery life of each charge cycle. These types of tasks include:

- Browsing the web.

- Watching a movie or TV show.

- Editing digital photos or video.

- Listening to music.

166

Click on the battery icon to access the Battery preferences options (see pages 167-169).

When you are using your MacBook you can always monitor how much battery power you currently have available. This is shown by the battery icon that appears at the top right on the Apple Menu bar. On battery power, the power level indicator is a solid color; on charge, it has a lightning symbol on it:

As the battery runs down, the monitor bar turns red as a warning:

Because of the vital role the battery plays in relation to your MacBook, it is important to try to conserve its power as much as possible. To do this:

- Where possible, use the mains adapter rather than the battery when using your MacBook.

- Use the Sleep function when you are not actively using your MacBook (see page 52).

- Use power management functions to save battery power (see next page).

Power management

To access power management options, click on the battery icon on the top menu bar. This shows the current power source and the apps that are currently using up the most energy. The battery preferences can be accessed by clicking on the **Battery Preferences...** link, or:

The Battery system preferences have been updated in macOS Big Sur (this was previously known as Energy Saver).

1 Access **System Preferences** and click on the **Battery** button

2 In the **Battery** window click on the **Usage History** button in the left-hand sidebar to view how the battery has been used, for the last 24 hours or 10 days

All of the Battery options display the current level of charge for the battery, at the top of the left-hand sidebar.

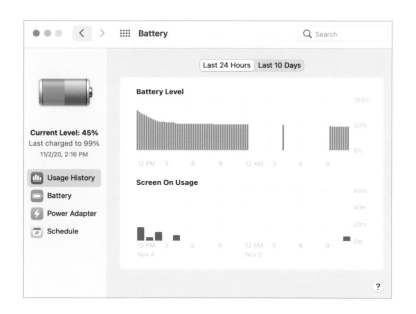

Managing the Battery

MacBooks have options for how the battery is managed within the Battery system preferences. These allow you to set options including individual power settings for the battery, and to view how much charge is left in the battery. To manage the battery:

1 Access **System Preferences** and click on the **Battery** button

2 Click on the **Battery** button in the left-hand sidebar

3 When the MacBook is on battery power, drag this slider to specify the period of inactivity after which the display is put to sleep

Hot tip

In the main Battery window, check On **Show battery status in menu bar** to make this visible on the top Menu bar on your MacBook.

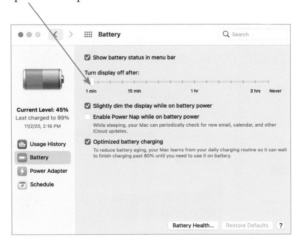

4 Check **On** these boxes to perform the required tasks

5 Click on the **Power Adapter** button in the left-hand sidebar

6 When the MacBook is connected with the power adapter, the options are similar. Drag the slider to specify when the display is put to sleep in the same way as for when the MacBook is on battery power. Select these buttons to prevent the MacBook from sleeping automatically when the display is off, allowing it to wake for network access, and enable it to wake for certain actions, such as checking for mail

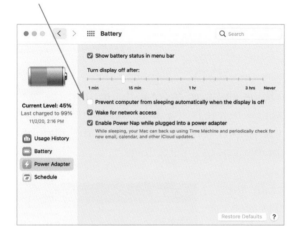

For the **Start up or wake** option in the Schedule window, click in the drop-down box to select from **Every Day**, a specific day of the week, **Weekdays** or **Weekend**.

7 Click on the **Schedule** button in the left-hand sidebar

8 Select settings if you want your MacBook to wake up and sleep at pre-defined times

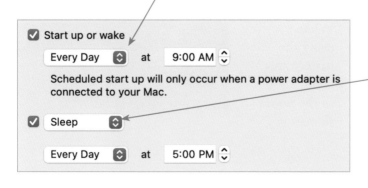

Click in this drop-down box to select an option for when the MacBook should **Sleep**, **Restart** or **Shut Down**.

Charging the Battery

MacBook batteries are charged using an AC/DC adapter, which can also be used to power the MacBook instead of the battery. If the MacBook is turned on and is being powered by the AC/DC adapter, the battery will be charged at the same time, although at a slower rate than if it is being charged when the MacBook is turned off.

The AC/DC adapter should be supplied with a new MacBook and consists of a cable and a power adapter. To charge a MacBook battery using an AC/DC adapter:

Don't forget

A MacBook battery can be charged whether the MacBook is turned on or off. It charges more quickly if the MacBook is not in use.

1 Connect the AC/DC adapter and the cable and plug it into the mains socket

2 Attach the AC/DC adapter to the MacBook and turn it on at the mains socket. When it is attached, this battery icon is displayed on the Apple Menu bar, including the amount of charge in the battery, as a percentage

40% 🔋

3 Click on the battery icon to view how long until the battery is charged, and the power source

40% 🔋 📶 Nick Vandome Q 🎚 🍎

Battery
Power Source: Power Adapter
1h 52m until fully charged

Using Significant Energy
⚙ System Preferences

Battery Preferences...

4 When using the battery, this is displayed as the power source

41% 🔋 📶 Nick Vandome

Battery
Power Source: Battery

Low Power and Dead Battery

No energy source lasts forever, and MacBook batteries are no exception to this rule. Over time, the battery will operate less efficiently until it will not be possible to charge the battery at all. With average usage, most MacBook batteries should last approximately five years, although they will start to lose performance before this and become less efficient. Some signs of a dead MacBook battery are:

- Nothing happens when the MacBook is turned on using just battery power.

- The MacBook shuts down immediately if it is being run on the AC/DC adapter and the cord is suddenly removed.

- The following window appears a few minutes or immediately after you have charged the battery and then started to use your MacBook on battery power:

When you are warned about a low battery, save all of your current work and either close down or switch to using an external AC/DC adapter for powering your MacBook.

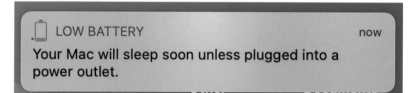

```
[🔋] LOW BATTERY                                    now
Your Mac will sleep soon unless plugged into a
power outlet.
```

Replacement battery

Although it is technically possible to change the battery of a MacBook it is not recommended, as it could damage the device and it would invalidate any warranty. MacBook batteries are designed to retain up to 80% of their original capacity for 1,000 complete charge cycles. However, if a MacBook battery does die, it can be replaced by using Apple's battery replacement service: details can be found on the Apple website at **https://support. apple.com/mac/repair/service**, or at your nearest Apple Store.

If a MacBook is still within its one-year warranty, a defective battery will be replaced for free, as long as it has not been damaged due to misuse.

Battery Troubleshooting

If you look after your MacBook battery well, it should provide you with several years of mobile computing power. However, there are some problems that may occur with the battery:

- It won't keep its charge even when connected to an AC/DC adapter. The battery is probably flat and should be replaced with a new one. (Even if the battery is flat, the MacBook will still operate using the AC/DC adapter.)

- It only charges up a limited amount. Over time, MacBook batteries become less efficient and so do not hold their charge so well. One way to try to improve this is to drain the battery completely before it is charged again.

- It keeps its charge but runs down quickly. This can be caused by using a lot of power-hungry applications on the MacBook. The more work the MacBook has to do to run applications, such as those involving videos or games, the more power will be required from the battery, and the faster it will run down.

- The MacBook only works when it is connected with the adapter. This is another sign that the battery is probably nearly flat and should be replaced.

- It is fully charged but does not appear to work. The battery may have become damaged in some way, such as from being in contact with water. If you know the battery is damaged in any way, take it to an Apple Store or a recognized Apple reseller and ask if the battery can be replaced. If the battery has been in contact with liquid, dry it out completely before trying to use it again. If it is thoroughly dry, it may work again.

- It gets very hot when in operation. This could be caused by a faulty battery, and it can be dangerous and lead to a fire. If in doubt, turn off the MacBook immediately and consult Apple. In some cases, faulty batteries can be recalled, so keep an eye on the Apple website to see if there are any details of this if you are concerned. Even in normal operation, MacBook batteries can feel quite warm. Get to know the normal temperature of your battery, so you can judge whether it is getting too hot or not.

Hot tip

If you are not going to be using your MacBook for an extended period of time, turn it off and store it in a safe, dry, cool place.

12 MacBook Maintenance

Despite its stability, macOS still benefits from a robust maintenance regime. This chapter looks at ways to keep macOS in top shape, ensure downloaded apps are as secure as possible, and some general troubleshooting.

Time Machine

Time Machine is a feature of macOS that gives you great peace of mind. In conjunction with an external hard drive it creates a backup of your whole system, including folders, files, apps, and even the macOS operating system itself.

Once it has been set up, Time Machine takes a backup every hour, and you can then go into Time Machine to restore any files that have been deleted or become corrupt since the last backup.

Setting up Time Machine

To use Time Machine, it first has to be set up. This involves attaching a hard drive to your MacBook via cable or wirelessly, depending on the type of hard drive you have. To set up Time Machine:

Make sure that you have an external hard drive that is larger than the contents of your MacBook, otherwise Time Machine will not be able to back it all up.

1 Click on the **Time Machine** icon in System Preferences

Time Machine

2 In the Time Machine window, click on the **Select Disk...** button

Select Disk...

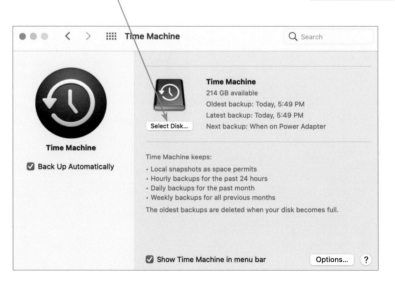

3 Connect an external hard drive and select it from the list

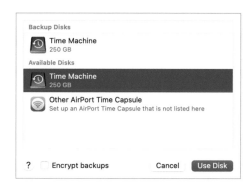

4 Click on the **Use Disk** button

Use Disk

5 In the Time Machine window, check **On** the **Back Up Automatically** option

Time Machine
☑ Back Up Automatically

6 The backup will begin. The initial backup copies your whole system and can take several hours. Subsequent hourly backups only look at items that have been changed since the previous backup

7 The record of backups and the schedule for the next one are shown here

When you first set up Time Machine it copies everything on your MacBook. Depending on the type of connection you have for your external drive, this could take several hours. Because of this, it is a good idea to have a hard drive with a USB-C or Thunderbolt connection to make it as fast as possible.

175

If you stop the initial backup before it has been completed, Time Machine will remember where it has stopped and resume the backup from this point.

...cont'd

Using Time Machine

Once Time Machine has been set up, it can then be used to go back in time to view items in an earlier state. To do this:

If you have deleted items before the initial setup of Time Machine, these will not be recoverable.

1. Access an item on your MacBook and delete it. In this example, the folder **Time Machine 1** has been deleted

2. Click on the **Time Machine** icon on the Dock or in the Launchpad

3. Time Machine displays the selected window in its current state (the **Time Machine 1** folder is deleted). Earlier versions are stacked behind it

Don't forget

The active item that you were viewing before you launched Time Machine is the one that is active in the Time Machine interface. You can select items from within the active window to view their contents.

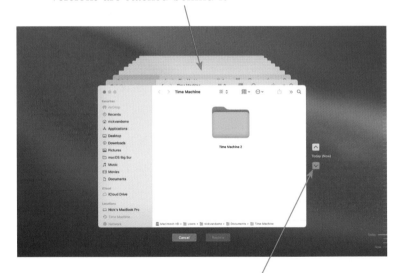

4. Click on the arrows to move through the open items, or select a time or date from the scale to the right of the arrows

5. Another way to move through Time Machine is to click on the pages behind the first one. This brings the selected item to the front

6 Move back through the windows to find the deleted item. Click on it and click on the **Restore** button to restore the item, in this case the **Time Machine 1** folder

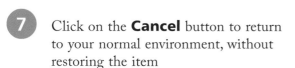

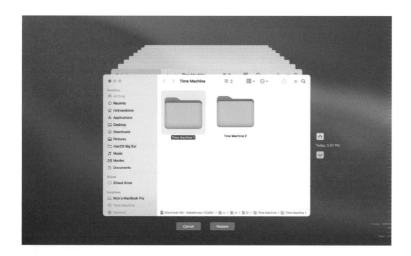

Items are restored from the Time Machine backup disk; i.e. the external hard drive.

7 Click on the **Cancel** button to return to your normal environment, without restoring the item

Cancel

8 The deleted folder **Time Machine 1** is now restored to its original location

Disk Utility is located within the **Applications** > **Utilities (Other)** folder.

Disk Utility

If there is a problem with a disk and macOS can fix it, the **Repair** button will be available. Click on this to enable Disk Utility to repair the problem.

Beware

If you erase data from a removable disk such as a flashdrive, you will not be able to retrieve it.

Disk Utility

Disk Utility is a utility app that allows you to perform certain testing and repair functions for macOS. It incorporates a variety of functions, and is a good option for both general maintenance and if your computer is not running as it should.

Each of the functions within Disk Utility can be applied to specific drives and volumes. However, it is not possible to use the macOS start-up disk within Disk Utility as this will be in operation to run the app, and Disk Utility cannot operate on a disk that has apps already running. To use Disk Utility:

Checking disks

1 Open Disk Utility then click the **First Aid** tab to check a disk

2 Select a disk and select one of the First Aid options

Erasing a disk
To erase all of the data on a disk or a volume:

1 Click on the **Erase** tab, and select a disk or a volume

2 Click **Erase** to erase the data on the selected disk or volume

Erase

System Information

This can be used to view how the different hardware and software elements on your MacBook are performing. To do this:

1 Open the **Utilities** folder and double-click on the **System Information** icon

2 Click on the **Hardware** link and click on an item of hardware

System Information is located within the **Applications** > **Utilities (Other)** folder.

3 Details about the item of hardware, and its performance, are displayed

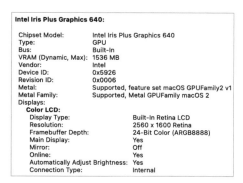

4 Similarly, click on network or software items to view their details

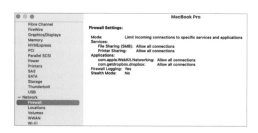

Activity Monitor

Activity Monitor is a utility app that can be used to view information about how much processing power and memory is being used to run apps. This can be useful to know if certain apps are running slowly or crashing frequently. To use Activity Monitor:

Activity Monitor is located within the **Applications** > **Utilities (Other)** folder.

Activity Monitor

1 Open Activity Monitor and click on the **CPU** tab to see how much processor capacity is being used up

System:	3.12%	CPU LOAD	Threads:	1,990
User:	3.94%		Processes:	594
Idle:	92.94%			

2 Click on the **Memory** tab to see how much system memory (RAM) is being used up

MEMORY PRESSURE	Physical Memory:	8.00 GB	App Memory:	3.20 GB
	Memory Used:	6.41 GB	Wired Memory:	2.09 GB
	Cached Files:	1.48 GB	Compressed:	1.13 GB
	Swap Used:	2.34 GB		

3 Click on the **Energy** tab to see details of battery usage (for a laptop)

ENERGY IMPACT		BATTERY (Last 12 hours)
	Remaining charge:	31%	
	Time remaining:	1:58	
	Time on battery:	65:10	

4 Click on the **Disk** tab to see how much space has been taken up on the hard drive

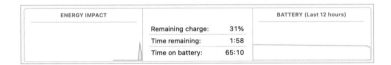

Reads in:	6,387,496	IO ⌄	Data read:	193.32 GB
Writes out:	3,117,316		Data written:	121.86 GB
Reads in/sec:	0		Data read/sec:	819 bytes
Writes out/sec:	34		Data written/sec:	207 KB

Updating Software

Apple periodically releases updates for its software; both its apps and the macOS operating system. All of these are now available through the App Store. To update software:

1 Open **System Preferences** and click on the **Software Update** icon

Software Update

2 Any available updates will be displayed. Click on the **Update Now** button to install an update

3 Click on the **Advanced...** button

Advanced...

4 Make the relevant selections for how updates are managed, including checking for them automatically and downloading new updates when they are available. Click on the **OK** button to confirm any changes

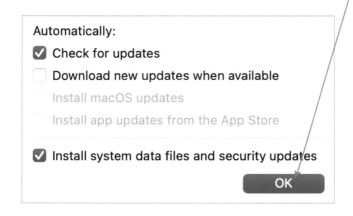

Hot tip

Software updates can also be accessed directly from the Apple menu button, located on the top Menu bar. If updates are available this is denoted on the **System Preferences** link on the Apple menu, or the **App Store** link.

Don't forget

For some software updates, such as those for macOS itself, you may have to restart your computer for them to take effect.

Gatekeeper

Internet security is an important issue for every computer user; no-one wants their computer to be infected with a virus or malicious software. Historically, Macs have been less prone to attack from viruses than Windows-based machines, but this does not mean Mac users can be complacent. With their increasing popularity there is now more temptation for virus writers to target them. macOS Big Sur recognizes this, and has taken steps to prevent attacks with the Gatekeeper function. To use this:

1 Open **System Preferences** and click on the **Security & Privacy** icon

Security & Privacy

2 Click on the **General** tab General

3 Click on one of these buttons to determine which location apps can be downloaded from. You can select from just the Apple App Store, or Apple App Store and identified developers, which gives you added security in terms of apps having been thoroughly checked

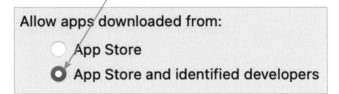

Allow apps downloaded from:
- App Store
- App Store and identified developers

4 Under the **General** tab there are also options for using a password when you log in to your account; if a password is required after sleep or if the screen saver is activated; showing a message when the screen is locked; or allowing your Apple Watch to unlock your MacBook (check **Off** the **Disable automatic login** option)

Privacy

Also within the Security & Privacy system preferences are options for activating a firewall and privacy settings. To access these:

1 Click on the **Firewall** tab

Firewall

2 Click on the **Turn On Firewall** button to activate this. Click on **Firewall Options** to change settings for the firewall

Firewall   Privacy

Turn On Firewall

Don't forget

A firewall is an application that aims to stop malicious software from accessing your computer.

3 Click on the **Privacy** tab

Privacy

4 Click on the **Location Services** link and check **On** the **Enable Location Services** option if you want relevant apps to be able to access your location

5 Click on the **Contacts** link and check **On** any relevant apps that are allowed to access your contacts

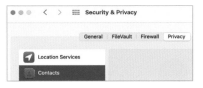

6 Click on the **Analytics & Improvements** link and check **On** or **Off** the options for sharing analytical information with Apple and app developers. This will include any problems, and helps Apple improve its software and apps. This information is collected anonymously

Problems with Apps

The simple answer

macOS is something of a rarity in the world of computing software: it claims to be remarkably stable, and it is. However, this is not to say that things do not sometimes go wrong, although this is considerably less frequent than with older Mac operating systems. Sometimes this will be due to problems within particular apps, and on occasion the problems may lie with macOS itself. If this does happen, the first course of action is to restart macOS using the **Apple menu > Restart** command. If this does not work, or you cannot access the Restart command as the MacBook has frozen, try turning off the power to the computer and then starting up again.

Force quitting

If a particular app is not responding, it can be closed down separately without the need to reboot the computer. To do this:

Beware

When there are updates to macOS, these can on rare occasions cause issues with some apps. However, these are usually fixed with subsequent patches and upgrades to macOS.

1 Select **Apple menu > Force Quit...** from the Menu bar

2 Select the app you want to close

3 Click the **Force Quit** button

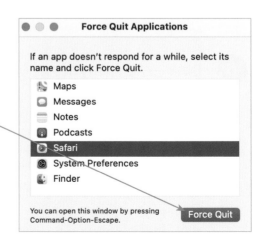

General Troubleshooting

It is true that things do occasionally go wrong with macOS, although probably with less regularity than with some other operating systems. If something does go wrong, there are a number of areas that you can check and also some steps you can take to ensure that you do not lose any important data if the worst-case scenario occurs, and your hard drive packs up completely:

- **Backup**. If everything does go wrong it is essential that you have taken preventative action in the form of making sure that all of your data is backed up and saved. This can be done with either the Time Machine app or by backing up manually by copying data to a flashdrive. Some content is also automatically backed up if you have iCloud activated.

- **Reboot**. One traditional reply by IT helpdesks is to reboot (i.e. turn off the computer and turn it back on again) and hope that the problem has resolved itself. In a lot of cases this simple operation does the trick, but it is not always a viable solution for major problems.

- **Check cables**. If the problem appears to be with a network connection or an externally-connected device, check that all cables are connected properly and have not become loose. If possible, make sure that all cables are tucked away so that they cannot be inadvertently pulled out.

- **Check network settings**. If your network or internet connections are not working, check the network settings in System Preferences. Sometimes when you make a change to one item, this can have an adverse effect on one of these settings. (If possible, lock the settings once you have applied them by clicking on the padlock icon in the Network preferences window.)

- **Check for viruses**. If your computer is infected with a virus this could affect the efficient running of the machine. Luckily, this is less of a problem for Macs as virus writers tend to concentrate their efforts toward Windows-based machines. However, this is changing as Macs become more popular, and there are plenty of Mac viruses out there. So, make sure your computer is protected by an app such as Norton AntiVirus, which is available from **www.norton.com**

In extreme cases, you will not be able to reboot your MacBook normally. If this happens, hold down the power button until the MacBook closes down. You should then be able to reboot it by turning it on in the usual way.

...cont'd

- **Check start-up items**. If you have set certain items to start automatically when your computer is turned on, this could cause certain conflicts within your machine. If this is the case, disable the items from launching during the booting-up of the computer. This can be done within the Users & Groups section of System Preferences by clicking on the **Login Items** tab, selecting the relevant item and pressing the minus button.

If you are having problems opening a document that you have been sent from a trusted source, contact them to make sure that the document has not been locked with a password. If you receive documents from someone you do not know, such as by email, do not open them as they may contain viruses or malware.

- **Check permissions**. If you or other users are having problems opening items, this could be because of the permissions that are set. To check these, select the item in the Finder, click on the **File** button on the top Menu bar and select **Get Info**. In the **Sharing & Permissions** section of the Info window you will be able to set the relevant permissions to allow other users, or yourself, to read, write or have no access.

Click here to view Permissions settings

Within the Screen Saver system preferences (Desktop & Screen Saver) is a button for **Hot Corners**. Click on this to access options for specifying what happens when the cursor is pointed over the four corners of the screen. This includes accessing the Notification Center and the Launchpad, or putting the MacBook display to sleep.

- **Eject external devices**. Sometimes external devices, such as flashdrives, can become temperamental and refuse to eject the disks within them, or even not show up on the Desktop or in the Finder at all. If this happens, you can try to eject the disk by pressing the trackpad button when the MacBook chimes are heard during the booting-up process.

- **Turn off your screen saver**. Screen savers can sometimes cause conflicts within your computer, particularly if they have been downloaded from an unreliable source. If this happens, change the screen saver within the **Screen Saver** tab in the **Desktop & Screen Saver** preference of System Preferences, or disable it altogether.

Index

Q

R

S